T0259669

Expert Apache Cassandra Administration

Sam R. Alapati

Apress®

Expert Apache Cassandra Administration

Sam R. Alapati
Flower Mound, Texas, USA

ISBN-13 (pbk): 978-1-4842-3125-8   ISBN-13 (electronic): 978-1-4842-3126-5
https://doi.org/10.1007/978-1-4842-3126-5

Library of Congress Control Number: 2017962948

Copyright © 2018 by Sam R. Alapati

This work is subject to copyright. All rights are reserved by the Publisher, whether the whole or part of the material is concerned, specifically the rights of translation, reprinting, reuse of illustrations, recitation, broadcasting, reproduction on microfilms or in any other physical way, and transmission or information storage and retrieval, electronic adaptation, computer software, or by similar or dissimilar methodology now known or hereafter developed.

Trademarked names, logos, and images may appear in this book. Rather than use a trademark symbol with every occurrence of a trademarked name, logo, or image we use the names, logos, and images only in an editorial fashion and to the benefit of the trademark owner, with no intention of infringement of the trademark.

The use in this publication of trade names, trademarks, service marks, and similar terms, even if they are not identified as such, is not to be taken as an expression of opinion as to whether or not they are subject to proprietary rights.

While the advice and information in this book are believed to be true and accurate at the date of publication, neither the authors nor the editors nor the publisher can accept any legal responsibility for any errors or omissions that may be made. The publisher makes no warranty, express or implied, with respect to the material contained herein.

Cover image designed by Freepik

Managing Director: Welmoed Spahr
Editorial Director: Todd Green
Acquisitions Editor: Jonathan Gennick
Development Editor: Laura Berendson
Technical Reviewer: Carlos Rolo
Coordinating Editor: Jill Balzano
Copy Editor: Mary Behr
Compositor: SPi Global
Indexer: SPi Global
Artist: SPi Global

Distributed to the book trade worldwide by Springer Science+Business Media New York, 233 Spring Street, 6th Floor, New York, NY 10013. Phone 1-800-SPRINGER, fax (201) 348-4505, e-mail orders-ny@springer-sbm.com, or visit www.springeronline.com. Apress Media, LLC is a California LLC and the sole member (owner) is Springer Science + Business Media Finance Inc (SSBM Finance Inc). SSBM Finance Inc is a **Delaware** corporation.

For information on translations, please e-mail rights@apress.com, or visit www.apress.com/rights-permissions.

Apress titles may be purchased in bulk for academic, corporate, or promotional use. eBook versions and licenses are also available for most titles. For more information, reference our Print and eBook Bulk Sales web page at www.apress.com/bulk-sales.

Any source code or other supplementary material referenced by the author in this book is available to readers on GitHub via the book's product page, located at www.apress.com/9781484231258. For more detailed information, please visit www.apress.com/source-code.

Printed on acid-free paper

I dedicate this book to the memory of Ibrahim Khan, a great cricketer and a greater human being.

The lessons I learned from "Khan Sahib" a long time ago still guide me.

Table of Contents

About the Author

 **Sam R. Alapati** is a Data Administrator at Solera Holdings, in Westlake, Texas, just outside Dallas. Before that, he was the Principal Big Data Administrator at Sabre, in Southlake, Texas. He has been working in the big data area for the past few years, after a long stint as an Oracle database administrator. Sam works with Apache Hadoop and Apache Spark, and now with Cassandra as part of his efforts to learn the role of NoSQL and other modern databases in the big data world. He has published several books on various areas of Oracle database administration, as well as a comprehensive handbook on administering Apache Hadoop.

About the Technical Reviewer

Carlos Rolo is a Datastax Certified Cassandra Architect, Cassandra MVP, and has deep expertise with distributed architecture technologies. He has become known and trusted by his peers for his ability to understand complex problems and for working well under pressure. He prides himself on being a tenacious problem solver while remaining a calm and positive presence on any team. In his spare time, he can be found playing water polo, enjoying his local community, or playing with his family and dog.

Acknowledgments

I wish to thank my good friend and longtime editor for many of my books, Jonathan Gennick. Jonathan encouraged me to keep going and supported my work when the going got tough during the writing of this book. Thank you, Jonathan, for your trust in me!

Carlos Rolo has been magnificent in many ways as the Technical Editor of this book. Carlos is well known for his mastery of the Cassandra database. I've benefited significantly from his painstaking technical reviews, as well as from his suggestions to clarify or improve several tricky technical concepts.

Jill Balzano, the Coordinating Editor, has been instrumental in ensuring that I stayed on target, with her usual combination of grace and efficiency. Jill's enormous patience and kindness helped make the writing and editing process a pleasurable one.

Amrita Stanley, Senior Executive Project Manager at SPi Content Solutions – SPi Global, was extremely helpful in making sure that things stayed on track. Thank you, Amrita, for being so very attentive to everything and helping me out with my requests.

Everything I do scholastically is due to my father's encouragement and nurturing. I thank my dear dad, the late Dr. Appa Rao, for his love and affection. I wish to thank my mom, Swarna Kumari, and my brothers, Hari Hara Prasad and Siva Sankara Prasad. I owe a round of thanks to all of my in-laws and the children of my brothers: Aruna, Vanaja, Teja, Ashwin, Aparna, and Soumya. I appreciate and thank my wife's indirect contribution to this book. I also am grateful for the support and encouragement of my twins, Nina and Nicholas, who brighten my life every single day!

—Sam R. Alapati

PART I

Introduction, Installation, and Configuration

CHAPTER 1

Apache Cassandra: An Introduction

Apache Cassandra is quite different from a typical relational database management system (RDBMS) such as MySQL or the Oracle database. This chapter sets the stage for the need for NoSQL databases in general, and for Cassandra.

In addition, this chapter explains the distinctive features of a Cassandra database and introduces several key concepts and terms that are unique to Cassandra.

You will learn about the key concepts that make Cassandra special. The chapter also introduces Cassandra's data modeling and explains key Cassandra terminology.

This book is about how to administer Cassandra, so I also explain the salient features of Cassandra administration, the tools that you'll use to manage Cassandra clusters, and the essentials of Cassandra database administration.

What Is Cassandra?

Apache Cassandra is a distributed NoSQL database system based on Amazon's Dynamo and Google's Bigtable. Cassandra is a fast, distributed database that's highly fault tolerant as well as scalable. It provides high availability and linear scalability, twin goals that traditional relational databases cannot satisfy when handling very large data sets.

Cloud applications require highly scalable back-end databases that are capable of distributed, massive workloads across clusters of servers. These applications require very fast access to data to satisfy interactive usage of the data stores by various applications, as well as ad-hoc queries. Cassandra is expressly designed for high-volume, low-latency cloud applications.

3

© Sam R. Alapati 2018
S. R. Alapati, *Expert Apache Cassandra Administration*, https://doi.org/10.1007/978-1-4842-3126-5_1

Facebook developed Cassandra for searching inboxes, and the project was open sourced in July 2008. The project was accepted into the Apache Incubator in March 2009, and it was made an Apache top-level project in February 2010. The name *Cassandra* derives from the ancient Greet prophet Cassandra.

In the following sections, I briefly describe the key features of Cassandra, which are

- Open source nature
- Fault tolerance
- High performance
- Distributed database architecture
- High scalability
- Elastic nature
- Ability to handle multiple types of data

Open Source Database

Cassandra isn't a commercial offering. It's a purely open source product, which means you not only don't pay for it, but you can also customize it for your needs. However, if you're working with production Cassandra clusters, it's important for you to be aware of DataStax, the "Cassandra company." DataStax is a company run by the main creators of the Cassandra database. DataStax offers commercial products such as the Cassandra DevCenter and Cassandra Enterprise to make it easy for you to set up and run Cassandra clusters. DataStax also offers several free courses in Cassandra data modeling, development, and operations, and is the leading evangelist for Cassandra through its documentation, courses, and conferences.

Fault Tolerance

Cassandra automatically replicates data to multiple nodes. This is in stark contrast to traditional relational databases, which require you to set up complex replication architectures to propagate data to multiple nodes for fault tolerance. For each update, you can choose either synchronous or asynchronous replication.

Since the database runs on a bunch of nodes that are similar, there's no single point of failure in a Cassandra cluster. Cassandra-based applications will survive the loss of even an entire Casandra datacenter (defined differently from the traditional data center of an organization), as you'll learn later in this book.

High Performance

Cassandra is a high performing database, and benchmarks have shown that its performance outstrips that of other NoSQL databases. In a major comprehensive NoSQL benchmark performed in 2012, Cassandra achieved the highest throughput for the maximum number of nodes in multiple experiments.

Cassandra provides a very high throughput, especially during writes. Predictable high performance means that you can meet strict business SLAs that require low latency.

Cassandra continues to deliver high performance despite an increase in the workload, or the addition of a large number of nodes to a cluster, due to the linear scalability of the database.

Distributed Configuration and High Resiliency

Cassandra installations use multiple nodes and there's no single point of failure. There is no master-slave relationship among the nodes; all nodes in a Cassandra cluster are identical.

Cassandra automatically replicates data to multiple nodes. You can locate the nodes in multiple racks, or different availability zones if you're using a cloud setup like Amazon Web Services. Spreading your data across multiple geographical regions will ensure that the data is always available.

Cassandra uses the concept of a datacenter, which is a set of nodes. A datacenter can be logical or physical, so you could have datacenters spread across the world. This feature, along with the replication of data, provides a very high degree of availability, as Cassandra can survive the loss of not only one or more nodes, but also that of an entire data center.

Flexibility

Cassandra uses an extremely flexible data model. A table can have a varying number of columns among its rows. Users can also change the schema at runtime. Cassandra's tunable consistency feature allows you to tune consistency at the read or write operation level. You can tune replication and read/write consistency guarantees based on whether your primary goal is reliability or speed.

Use of Commodity Servers

Unlike relational databases, most of which require highly robust and expensive hardware, Cassandra can run on commodity hardware. Commodity hardware is far cheaper than the high-grade servers that host databases such as Oracle. Although more prone to failures, the commodity servers work very well since Cassandra's architecture doesn't involve a single point of failure.

Ease of Management

Cassandra is easy to manage due to its simple architecture and the lack of too many moving parts. Oracle and other relational databases have many features, which leads to database bloat, which could contribute to the instability of the database.

Administering a Cassandra database, especially through a framework such as DataStax, is simple compared to managing a complex database such as an Oracle database. Backing up and recovering the database, tuning SQL queries, and managing replication are all complex tasks in an Oracle database. Cassandra, by keeping things simple, makes it easy for you to get up to speed.

A key thing to remember when you scale a Cassandra cluster by adding more nodes is that you don't need to add more members to the team that manages Cassandra. The same team that manages a 20-node cluster can handle a 1,000-node cluster. Therefore, the cost of management goes down as you scale a Cassandra database.

Highly Scalable

Cassandra is highly scalable. To analyze more data, you just add more nodes to your deployment. Cassandra offers linear scalability, which means that if you need to handle a threefold increase in writes, you increase your node count by three.

You can scale Cassandra easily, without any downtime or interruption of ongoing work.

Large Cassandra production deployments include Netflix, with over 2,500 nodes and 420TB of data, and Apple, with over 75,000 nodes and over 100PB of data.

One of the keys to Cassandra's high performance is that it is horizontally scalable. Unlike vertical scaling strategies that depend on hiking a server's processing, storage, and memory capabilities, horizontal scaling adds more nodes to a cluster.

As you add new machines to a Cassandra cluster, both read and write throughput increase linearly; that is, you can double your throughput, for example, by doubling the number of machines in the cluster. In addition, you don't pay the cost of service interruptions because the database remains available and your applications continue to run unimpeded during the expansion of the cluster's size.

It's Good Old SQL!

Finally, the Cassandra Query Language (CQL) is similar to the SQL language constructs most of us are already familiar with, making it quite easy for developers and administrators to get up to par with CQL.

The easiest way to interact with a Cassandra database is through the CQL shell, invoked by the utility cqlsh. CQL uses a syntax similar to regular SQL. You can perform data manipulation (DML) and data definition language (DDL) tasks through cqlsh.

Problems Posed by Relational Databases

You run into various problems with relational databases when you attempt to scale to very large data volumes. The notion of high availability that's offered by relational databases is also questionable at times.

When you are analyzing data using Linux utilities like sed and awk or simple Python scripts, most of the time you don't even need a database. More often, you'll be dealing with *medium data* where you use a database such as Mysql, Oracle, PostgreSQL, or SQL Server. These databases offer the well-known ACID guarantees.

As your data sizes grow, you can vertically grow these relational databases by adding CPUs that are more powerful and more RAM, but there's a finite limit to scalability with these relational databases, and the expense gets out of hand because the databases aren't distributed.

RDMSs and Big Data

Relational databases often employ replication in a master-slave format to solve the problem of read-heavy workloads. When you add replication to a relational database such as MySQL, the ACID guarantees become hard to obtain. When you replicate data in MYSQL, for example, there is usually a replication lag between the master and the slave. If a client reads from the slave before the slave is replicated with the data, the guarantee of consistency (the C in ACID) is lost.

Performance Problems Due to the Third Normal Form

In relational databases, it's common to place data in the well-known third normal form. The third normal form has several consequences for query performance, making it hard for you to scale. Queries are increasingly complex, with numerous joins and resource-hungry grouping operations such as ORDER BY and UNION.

The queries are adhoc and users want immediate results. If your datasets are larger than the memory you allocate to the database, results are going to be slow due to the heavy amount of disk seeks.

One way out of this performance predicament is to create multiple denormalized tables at read time, so the reads don't involve too many expensive joins. However, this means you're going to end up with duplicate sets of data, which violates the third normal form.

Problems Due to Sharding

Sharding is where, instead of having one database, you split up the data among multiple databases.

Sharding means you denormalize everything and that you can't do any more aggregations or joins. Your data is spread over numerous machines, and aggregations require you to touch every shard. Adding the shards requires you to move the data. Since sharding inhibits performance, you're forced to denormalize, meaning you keep multiple copies of key data that you query often, such as one copy based on Customer Id and the other by State.

Schema changes are hard when you have multiple shards, since you must propagate the changes to all the shards. In addition, if you want to double or triple the size of your database, you'll need to re-shard all the data.

The Myth of High Availability

High availability is a major benefit that users of replicated databases often seek. High availability isn't simple, however. High availability involves numerous moving parts and imposes a heavy overhead. Despite all the costs, high availability is very hard to achieve because master-slave environments do suffer frequent downtimes caused by configuration changes, drive and power supply failures, and so on.

In addition to the unplanned downtime, you should also deal with planned downtimes due to OS patches and upgrades. Failover processes, whether they're automatic or manual, aren't error free, and often involve a delay in bringing up the slave or standby database, resulting in downtime.

Managing relational databases such as MySQL running over multiple data centers isn't easy; it requires constant attention from a team of highly skilled database professionals and Linux engineers.

Cassandra simplifies your database architecture by offering the following features:

- *Peer-to-Peer Architecture*: Cassandra uses a peer-to-peer architecture, with no concept of master and slave instances. Doing away with the master/slave concept means that there are fewer moving parts. This strategy leads to a simple architecture where all nodes are equal.

- *No Failovers*: The peer-to-peer architecture means that there's no concept of a failover. The database doesn't have to deal with complex state maintenance operations such as the election of a leader.

- *Reliability*: The database fully expects failure, and failure of individual components in the architecture is never going to be a showstopper. Your users are completely oblivious to the fact that one or more nodes, or even an entire data center, have crashed or are unavailable for other reasons.

- *Data Locality*: Instead of gathering data that's scattered over a large number of nodes, Cassandra aims for data locality.

- *Lower Cost*: Cassandra reduces the cost of scaling (scaling with a traditional relational database can be expensive). Since Cassandra runs on commodity servers, your cost is much lower than scaling up a relational database with powerful servers. Instead of purchasing a few servers that cost hundreds of thousands of dollars, you acquire hundreds of servers that cost a few thousand bucks each.

Cassandra is a leading member of the NoSQL database constellation, so it's a good idea to review the need for NOSQL databases, the various types of NoSQL databases, and the types of problems they solve.

The Rise of NoSQL Databases

Just as relational databases arose to overcome the problems of data anomalies and the difficulty of reusing the same database for new applications, NoSQL databases have reasons for their emergence as a major player in the big data world. The failure of the relational model to scale so they can meet the growing demands for huge volumes of read and write operations led to the emergence of the NoSQL databases.

NoSQL databases can run on a single server or on multiple servers; they are referred to as *distributed databases* when they run on multiple servers. Most NoSQL databases handle large data sets and thus are distributed databases.

Unlike in a relational database design, where the model is everything and the structure of the entities and the relationships among them drive the design of the database, performance is the mantra in a NoSQL database. You'll still have entities and relationships, but preservation of the relationships isn't sacrosanct; the goal is performance.

There are several types of NoSQL databases, each geared to solving a different type of problem. The following sections briefly describe the main types of NoSQL databases.

Key-Value Databases

Key-value databases employ a simple data model based on sets of keys and values. The keys are the identifiers with which you look up data, and the values are the data that is associated with the keys.

The simple model employed by key-value databases means that there are no tables and related entities such as columns and constraints. These databases don't support the use of SQL to query the database.

Amazon's DynamoDB, Riak, Oracle BerkeleyDB, as well as well as caching databases such as Redis and MemcacheD are all key-value databases.

Document Databases

Document databases store documents, which are collections of data items that you store in a flexible structure. A document is simply a data structure that the database stores as strings or a binary representation of strings. Documents in this context are semi-structured entities in a standard format such as JavaScript Object Notation (JSON) or Extensible Markup Language (XML).

Document databases can store multiple attributes in a single document instead of storing each attribute of an entity with a separate key, as in a key-value database.

As with key-value databases, document databases use identifiers, but their values, which the database stores as documents, are normally more complex than the simple keys in a key-value database.

You don't need to define a schema before you add data to a document database. Adding a document will create the necessary data structures to support the document. Since there's no fixed schema, developers have more flexibility with these databases as compared to relational databases.

Popular document databases include MongoDB, CouchDB, and CouchBase.

Column Family Databases

Column family databases borrow some of the characteristics of relational databases by organizing data into collections of columns. However, they trade off some of the essential features of relational databases, such as the ability to join tables, for the sake of enhanced performance.

The rows in a column family database can have different columns. When you're dealing with a large number of columns, you can group columns into sets of related columns. These collections of related columns are called *column families*. It isn't unusual for a column family database to contain millions of columns.

As with a document database, a column family database doesn't require you to have a predefined, fixed schema.

Typically, a column family database is denormalized, and you find all information about an object in a single, wide row. A row in Cassandra can have as many as two billion columns. The query languages you use to work with column family databases are similar to SQL, and they use SQL-like statements such as SELECT, UPDATE, and INSERT.

Apache Cassandra and Apache Hbase (usually bundled with Hadoop distributions) are two well-known column family databases.

Graph Databases

A graph database is a specialized NoSQL database that uses structures called *nodes* and *relationships* (called *vertices* and *edges* in formal terms) instead of modeling data with columns and rows. Graph databases are designed for modeling objects and the relationships among the objects.

11

Graph databases have nothing to do with the visualization aids you normally refer to as a graph (or chart). They derive their name from graph theory, a branch of mathematics that studies objects by representing them as vertices and the relationships among them as edges.

Graph databases model the adjacency between objects. A node in the graph database contains pointers to objects that are adjacent to it. The pointers enable fast operations that require passing through the paths in a graph.

A node in a graph database is an object with an identifier and a set of attributes. A relationship between two nodes is a link that contains attributes about that relationship.

In models that involve multiple paths between nodes, graph databases enable you to query data efficiently. Querying this type of data is much harder using a tabular representation of data since those formats require recursive SQL constructs such as Oracle's CONNECT BY clause to find the paths.

Graph databases can represent a wide range of entities in the real world. For example, a city can be a node, and the information about the distance between the cities and the travel times between the cities can be stored in the relationships between the cities.

Neo4j is a popular graph database.

Special Capabilities of Cassandra

In the following sections, I highlight some of the special capabilities of Cassandra, which set it apart from other NoSQL databases. By no means are these capabilities exclusive to Cassandra, as some of the alternatives do offer some of these capabilities, but no other database comes close in terms of offering all of these features.

Distributed and Decentralized Database

Cassandra is fully distributed. Each machine in a Cassandra cluster takes care of an equal amount of activity in terms of data processing. In a big data framework such as Hadoop, things are quite different. There's a master node called the NameNode, which manages the reading and writing of all data. Similarly, MongoDB has its mongos, which require special treatment when compared to regular nodes, since they could potentially be a bottleneck.

Installation and troubleshooting are much easier in a Cassandra cluster when compared with other clusters, since all nodes are identical in terms of their functions. Cassandra's peer-to-peer strategy means that adding capacity is a trivial exercise.

You can scale a database such as MySQL by setting up some nodes as master nodes and the other nodes as slaves. Cassandra is designed as a decentralized database, meaning that all nodes are the same; there's no concept of master/slave nodes. This also means that there is no single point of failure since there are no special hosts, and the cluster continues operations regardless of node failures.

Log-Structured Storage Engine

Cassandra uses an efficient log–structured engine that turns updates into sequential I/O.

Cassandra's storage engine doesn't ever require or read existing data. It only appends updated data, thus making updates very fast.

Locally Managed Storage

Cassandra manages all its storage locally. This may seem trivial at first, but when compared to NoSQL databases such as Hbase, which rely on an underlying storage system such as HDFS (Hadoop Distributed File System) for replication, Cassandra offers many more capabilities.

You can use the locally managed storage strategy to place Cassandra's commitlog on a separate disk. You can also mix SSDs and HDDs in the same cluster.

You can reduce the I/O based on access patterns to reduce the cost of using extremely fast storage. DataStax offers DES tiered storage, which allows you to convert the reduced I/O into cost savings on the storage side.

DSE can transparently move aged data from the fast SSD tier to a slower storage tier. Both time series data and social interaction data are good examples of the types of data that can benefit from DSE tiered storage.

Ability to Handle Multiple Types of Data

Cassandra handles various types of data, such as structured, semi-structured, and unstructured data.

Row-Oriented

Although Cassandra falls under the column-oriented database type, which stores its data storage in columns, it actually is a partitioned data store, with "partitioned" referring to the fact that the database uses unique keys for each row to distribute the rows across multiple nodes. It stores data in sparse hash tables, with "sparse" alluding to the fact that all rows may not have the same columns.

One of the most interesting and notable features of Cassandra is tunable consistency, and I dedicate an entire section to a discussion of database consistency since it's critical that you understand this topic well.

Consistency, ACID Requirements, the CAP Theorem, and BASE

Consistency is a database property that ensures that a read always retrieves the most up-to-date value. All nodes that have the same data should have a consistent state of a write.

Unlike relational databases, which offer strict consistency to ensure transactions are always correct, NoSQL databases offer something called eventual consistency.

In NOSQL databases, there's a tradeoff between consistency and availability. Cassandra trades some consistency to provide high availability. To be more accurate, one should say that Cassandra offers tunable consistency, which means you can choose the level of consistency that's best for you and compromise accordingly on the availability front.

To understand tunable consistency, a key feature of Cassandra, it's useful to delve into the basics of the traditional ACID requirements for relational databases, Brewer's CAP theorem, and the BASE properties.

The ACID Requirements

As with other document databases and other databases that fall under the broad umbrella of NoSQL databases, Cassandra doesn't support the well-known ACID properties offered by relational databases. ACID properties are the hallmark of all modern transactional databases (such as MySQL and Oracle) and refer to the following set of principles:

- *Atomicity*: This is the all-or-none principle that requires that if even one element of a transaction fails, the entire transaction will fail. The transaction succeeds only if the database successfully performs all its tasks.

- *Consistency*: This property ensures that the database fully completes all transactions, by requiring that the database must be in a consistent state both at the beginning and at the end of a transaction.

- *Isolation*: Each transaction should be independent of the other transactions. No transaction must have has access to any other transaction that is in an unfinished state.

- *Durability*: Once a transaction completes, it's "permanent." That is, the database records the transaction in persistent storage, ensuring that the transaction survives a system breakdown such as a power or disk failure.

While the ACID requirements served traditional relational databases just fine for many years, the popularity of non-relational data such as unstructured data, non-relational data, and the proliferation of distributed computing systems led to new views about the required transaction properties that databases must satisfy.

The CAP Theorem

The Consistency, Availability, and Partition Tolerance (CAP) theorem sought to refine the requirements that you need to meet for implementing applications in modern distributed computing systems. The CAP theorem stands for the following principles:

- *Consistency*: This is the same as the ACID consistency property. Satisfying this requirement means that all clients of a data store get responses that "make sense." If Client A writes 1 followed by 2 to node X, a different client, Client B, can't read 2 first and then 1.

- *Availability*: The system must be available when requested. All operations on the data store must eventually return successfully.

- *Partition Tolerance*: If the network fails to transmit messages between two sets of nodes, the system must continue to work correctly. The failure of a single node in a distributed system must not lead to the failure of the entire system. The system must be available even if there's some data loss or a partial system failure.

The problem is, at any given time, a distributed system can usually support only two out of the three requirements listed here. This means that tradeoffs are usually inevitable when using distributed data stores such as NoSQL databases. That is why we have the famous statement "Consistency, availability, partition tolerance. Pick two."

Most modern distributed data stores such as Cassandra offer availability and partition tolerance, sacrificing consistency. The idea here is that short periods of application misbehavior are preferable to short periods where the database is unavailable.

If two nodes can't communicate, and you require strict consistency, the system will appear to be down. A database such as Casandra chooses to be highly available during network partition (when the network drops messages between nodes), instead of satisfying the consistency principle.

Strict consistency, especially when you have data centers that are distributed across the world, is something you may never be able to satisfy, due to the unavoidable latency among the data centers. Availability is a more practical principle that a distributed database can try to satisfy.

For reliability purposes, meeting the availability and partition tolerance requirements is essential, of course. That means that the consistency requirement is often at risk. However, leading NoSQL databases such as Cassandra and Amazon's DynamoDB deal with the loss of consistency just fine. How so? This is possible due the adoption by these databases of something called the BASE system, which is a modified set of ACID requirements to fit modern NoSQL and related non-relational databases. Here's what BASE stands for:

- *Basically Available*: The system guarantees the availability of data in the sense that it'll respond to any request. However, the response could be a "failure" to obtain the request data set, or the data set returned may be in an inconsistent or changing state.

- *Soft*: The state of the system is always "soft" in the sense that the "eventual consistency" (the final requirement) may be causing changes in the system state at any given time.

- *Eventually Consistent*: The system will eventually become consistent once it stops receiving new data inputs. As long as the system is receiving inputs, it doesn't check for the consistency of each transaction before it moves to the next transaction.

Amazon's DynamoDB, which lies behind Amazon's shopping cart technology, stresses high availability, meaning it can afford to go easy on the consistency angle. These types of databases eschew the complex queries necessary to support consistency in the traditional sense, settling instead for the eventual consistency goal.

Eventual consistency in this context means that in a distributed system, not all nodes see the same version of the data; at any given time the state may diverge between nodes. That is, it's possible for some nodes to serve stale data. However, given sufficient time, the state will come to be the same across the system.

MongoDB, a popular NoSQL database, on the other hand, favors consistency and partition tolerance over high availability.

The discussion on CAP and BASE shows that while NoSQL databases such as Cassandra have their advantages, particularly in the way they support horizontal scaling and the efficient processing of non-relational data, they do come with unique drawbacks and involve crucial sacrifices in terms of simultaneous support for traditional principles such as data consistency and availability.

Consistency, while it's a laudable objective, has a negative impact on cost-effective horizontal scaling. If the database needs to check the consistency of every transaction continuously, a database with billions of transactions will incur a significant cost to perform all the checks. The idea of consistency is not practical in a large distributed database.

It's the principle of eventual consistency that has allowed Google, Twitter and Amazon, among others, to interact with millions of their global customers, keeping their systems available by supporting partition tolerance. Without the principle of eventual consistency, there wouldn't exist all these systems today that deal successfully with the exponential rise of data volumes due to cloud computing, social networking, and related modern trends.

Note Cassandra lets you set a consistency level that clients must specify on all operations. This way, you can determine how many nodes in the Cassandra cluster must acknowledge a write operation for the database to consider the write successful.

Let's say you set the replication factor is 3 for some data, meaning that the databases must propagate all data to three different nodes. By setting your consistency level to 2, for example, you tell Cassandra to deem an update or a write as successful even if some nodes are down.

Cassandra's Drawbacks

Although Cassandra offers numerous benefits, you ought to recognize that there are some strong drawbacks as well to the database, especially when you're coming from a relational database background.

In the following sections, I review Cassandra's drawbacks from two viewpoints: querying and writing data.

Problems with Querying Data

The chief drawbacks of Cassandra when querying data are the fact that it's an eventually consistent database and that it doesn't offer support for standard database joins.

Eventual Consistency

Cassandra automatically replicates data across the nodes of a cluster. There's no master node, the data propagation does involve some latency, and therefore, you'll have eventual consistency.

As explained earlier, eventual consistency means that in a distributed setup, to achieve high availability there's an informal guarantee that if no new updates are made to data items, eventually all accesses to that data item will return an identical (last updated) value. You can mitigate the side effects of eventual consistency by using *quorum reads and writes*.

No Joins

Cassandra doesn't use table joins, which are a key feature of relational database queries. You therefore also don't have any foreign keys.

You get around the lack of transactions by denormalizing your data, and use duplicated data and a query-oriented storage format. Another strategy is to use powerful data analysis tools such as Apache Spark together with Cassandra.

No Indexes

Relational databases employ both a primary key that uniquely identifies rows, as well as multiple other indexes called *secondary indexes* to speed up retrieval. Cassandra tables do use a primary key, but the database has limited support for secondary indexes.

In Cassandra, you'll find few cases that call for the creation of a secondary index. Indiscriminate use of secondary indexes hurts performance.

Problems with Writing Data

Cassandra also suffers from some drawbacks when writing data due to its lack of support for traditional transactions and the overhead due to immutable tables and mutation.

Lack of Support for Transactions

Transactions are a huge part of traditional relational databases, but Cassandra, which doesn't support transaction-related concepts such as rollbacks or locking mechanisms that are so critical when dealing with relational databases, doesn't support them.

Cassandra does support lightweight transactions, but they are expensive. Cassandra uses the Paxos protocol to implement lightweight transactions to handle concurrent operations that read and write data items in a sequential order. The Paxos protocol is implemented by a serial consistency with real-time constraints, ending up with a transaction isolation that's similar to that offered by a serialized isolation level in a database such as Oracle. Chapter 7 discuss the Paxos algorithm and linearizable consistency in detail.

Overhead Due to Immutable Tables and Mutation

Cassandra stores its data in immutable structures called SSTables. When you update data, Cassandra spreads the data across multiple SSTables. When you delete data, it creates markers called tombstones to ensure that the database correctly deletes data across all nodes of the cluster.

Since a read operation must gather data from multiple tables, there is inherent overhead for reads. As you will learn in Chapter 11, cleaning up SSTables by compacting them ameliorates this overhead.

Who Should Use Cassandra?

I have explained the key drawbacks of Cassandra. The drawbacks should not deter you from using Cassandra; far from it! The drawbacks should serve to tell you if you ought to use Cassandra for your use cases.

Cassandra is an excellent choice when you're dealing with data that doesn't change (immutable data), and where updates and deletes are rare. This means that Cassandra is great for handling huge volumes of immutable data from personalization, fraud detection, time series, and sensor data.

Cassandra is also an appropriate choice if you're primarily looking for a database that performs fast writes.

Cassandra Optimizations

While relational databases widely use strategies such as query optimization and compression, Cassandra uses various optimizations, some of which you usually won't find in relational databases. The following sections describe key Cassandra performance optimizations.

Data Caches

Cassandra offers integrated caching and distributes the cached data across the cluster. When a node becomes unavailable, the client reads the data from another cached copy (replica) of the data. There's no separate database caching tier. Cassandra offers two types of data caching: a partition key cache and row caching.

- The partition key cache caches the partition index for a Cassandra table to reduce seek times.

- Row caching will cache an entire row. When the number of reads is much larger (about 95%) than the number of writes, Cassandra recommends that you use row caching.

Compression

Cassandra allows you to compress data to reduce the volume of data stored on disk, as well as lower the disk I/O. Compression is usually expensive in a relational database, due to the additional CPU cycles and disk I/O for compressing and decompressing data. Since Cassandra tables are immutable (once they are flushed to disk, they aren't ever modified), the database doesn't need a recompression cycle to process writes. This means that Cassandra compresses the SSTables just once, when it initially writes the tables to disk from the memtables.

In addition to SSTables, it's also possible to compress Cassandra's commit log to improve write performance.

Cassandra offers multiple compaction strategies, and I discuss them in Chapter 11.

Bloom Filters

Cassandra uses bloom filters to determine if a table (SSTable) has data for a specific partition. The idea is to eliminate the reading of whole SSTables when searching for specific bits of data. Cassandra uses bloom filters for index scans, but not for range scans.

Compaction

SSTables are immutable, meaning Cassandra doesn't overwrite existing data with inserts or updates. Instead, Cassandra writes a time-stamped version of the inserted or updated data in a new SSTable. As a result, over time, Cassandra ends up with multiple versions of a row in different SSTables.

Cassandra also doesn't perform deletes by immediately removing the data you delete. It simply marks the data with a *tombstone*, indicating that the data is deleted data.

Cassandra periodically merges the SSTables to discard old data, a process called *compaction*. Compactions results in the assembling of a single complete row using the latest version of each of that row's columns.

There are various compaction strategies you can use, the choice depending on whether your workload is write- or read-intensive.

Is Cassandra Appropriate for You?

Cassandra is not a general-purpose database! The following are the common use cases for which Cassandra will be ideal.

- *Large Environments*: Cassandra is a database for high-volume transactions. A good RDBMS can handle a large number of transactions and can do so while running on just a single server or a couple of them. Your use case must involve super-heavy throughout requirements and high traffic, and a requirement for low-latency work, for you to consider using Cassandra.

- *Write-Heavy Workloads*: Cassandra is well suited for use cases that involve heavy writes, since it offers high throughput for write operations. Cassandra is an excellent choice if you have numerous clients and high write volumes.

- *Distributed Data*: Since Cassandra works well with geographically distributed clusters, a global application that benefits by storing data near the user can potentially use Cassandra.

Cassandra isn't the best fit for all use cases. However, the reasons described earlier in this chapter, such as fast performance and support for a large number of complex data types, make it an outstanding excellent tool for highly scalable storage.

You can continue to use MySQL or Oracle for most of your storage needs. If a business unit sees a need for storing very large volumes of data, that's where you want to use Cassandra. It makes sense to use Cassandra in these cases rather than trying to stretch the capacities of the relational databases to manage and analyze huge troves of data.

Cassandra Data Modeling Essentials

Chapter 4 focuses on Cassandra data modeling and several other chapters throughout this book elaborate on the key Cassandra data modeling concepts. In this introductory chapter, I would like to mention the key principles or rules for successful data modeling in a Cassandra database.

You need a new way of thinking to write applications that scale on a distributed database. The first thing to remember is that Cassandra isn't a drop-in replacement for a relational database. This means that you can't expect to drop your data and queries from a relational database into Cassandra and expect it all to work fine. To get the most out of Cassandra, you should design your application using Cassandra's data modeling principles and rules.

Cassandra uses a data model based on the expected queries that you'll run against the database, and not on modeling the entities and the relationships among them.

It's common for users transitioning to a distributed database such as Cassandra to try to use materialized views, indexes, and user-defined functions to scale their applications. When you use these traditional strategies in a Cassandra-based application, you'll be in for quite a surprise.

With this background, let's review the key principles behind Cassandra data modeling.

Structure Your Data by the Queries

Since Cassandra is very efficient in writing data, you incur minimal overhead for denormalizing data and duplicating it. Duplicating data is going to make your reads faster. Ideally, a read should be able to get all its data from a single partition.

Avoid Updates and Deletes

By its design, Cassandra works well with an append-only, immutable set of data. You'll pay a stiff overhead for clearing mutations and deletes from the "ring."

Evenly Distribute Your Data

You must ensure that you spread your data evenly across a Cassandra ring, so a single node isn't stuck with processing a disproportionately large amount of data. Appropriate selection of the partition keys is critical to ensuring an even data distribution.

Avoid Querying Across Partitions

It takes longer for a read to process and aggregate data from multiple nodes. Partitioning strategy and location is critical here because querying across partitions may result in a heavy overhead.

What a Cassandra Administrator Ought to Know

This book is primarily about administering Cassandra, although it'll be of significant use to Cassandra architects and developers as well.

As with any database, Cassandra requires the tuning of various configuration parameters, selection of the right data model, and other strategies to produce peak performance. While several of the skills that you've acquired in managing a relational database are readily applicable to the Cassandra database, you'll need to learn additional things to be a successful Cassandra database administrator.

Traditional skills such as optimizing the operating system through configuring the kernel, monitoring the operating system through various tools, and SQL query optimization techniques are all going to be helpful to you when administering a Cassandra database. In addition to these skills, you'll need to focus on the tools and skills I describe in the following sections.

Cassandra Tools

Cassandra offers several useful administration tools to monitor and manage your clusters. The following is a brief description of the key tools you'll often use.

- *The cassandra Utility*: The cassandra utility enables you to start a Cassandra instance. In addition, the utility enables you to perform other tasks, such as replacing a dead node.

- *The SSTable Utilities*: Cassandra offers about a dozen useful utilities that enable you to perform SSTable tasks such as dumping the contents of a table, printing a table's metadata, splitting a table into multiple tables, and listing the SSTable files for a table.

- *The nodetool utility*: The nodetool utility is a command line tool to monitor and manage a cluster. You'll start using nodetool in Chapter 2, and by the time you reach the end of the book, you'll have learned every single nodetool command option.

- *The cassandra-stress tool*: This is a stress testing utility you can use to benchmark and load test your clusters. The tool is especially useful in understanding the scalability of your database and optimizing your data models.

Acquiring Parallel Remote Tools

Cassandra is a distributed database, and a cluster can often use dozens and even thousands of nodes. For those of you coming from a single-node-database relational database world, this will probably come as a major surprise.

It's a good idea for you to learn to use a distributed shell tool. There are several distributed SSH tools, such as parallel SSH and Cluster SSH, that enable SSH access to multiple nodes.

You will also benefit from using a tool such as pdsh (parallel distributed shell) to simultaneously move files and run commands on multiple nodes.

If you're dealing with more than a handful of nodes (say more than 8 or 10), it's a good idea to use a tool such as Chef or Ansible to perform many operations.

Understanding Failure Scenarios in a Cassandra Database

Losing a disk is somewhat of a rare event when managing traditional databases, since most of them run on enterprise-grade hardware. Cassandra, as all big data, runs on commodity hardware. Experiencing a disk or even a node failure isn't a rare event in the big data world.

A Cassandra administrator needs to know how to handle node failures and how to expand the node capacity by adding nodes on demand.

When operations fail due to updates and writes not being synchronized correctly across the cluster, administrators run a "repair" operation, an operation that's unique to Cassandra.

Everything in Cassandra is Java-based, so knowing how to read Java error stacks and trace files is critical. Learning about common Java exceptions, such as out-of-memory (OOM) errors and null pointer errors, helps.

Monitoring and Alerting

You can monitor and manage Cassandra, which is based on the Java language, through the Java Management Extensions (JMX).

It's more common to manage and monitor a Cassandra cluster through the nodetool utility offered by Cassandra. Nodetool is JMX-compliant, and you can use it to get various Cassandra metrics and to perform common tasks such as adding and decommissioning nodes. You can also use other JMX tools such as JConsole to perform various tasks.

Tuning the Java Virtual Machine

Cassandra is written in Java, and everything in Cassandra runs inside a Java Virtual Machine (JVM). Traditional database administration really doesn't require a knowledge of a JVM. However, that knowledge is critical when dealing with Cassandra databases. Chapter 11 introduces the concepts underlying JVM tuning and how to manage garbage collection (the most important area of tuning a JVM).

In the following section, I introduce the architecture of Cassandra plus several interesting concepts and key terms that you'll encounter during the rest of the book.

A Quick Introduction to the Architecture of Cassandra

Cassandra is a distributed database system. It distributes data among multiple nodes organized as a cluster to ensure that data is accessible even when there are system and hardware failures.

The nodes in a Cassandra cluster use a peer-to-peer communication protocol to exchange their state information.

Cassandra table rows are stored in tables, each with a mandatory primary key. A keyspace is a logical entity to group a set of tables, usually belonging to a single application. You access the data stored in the tables via the CQL language constructs, which use a similar syntax to that of SQL. You can access CQL through cqlsh and via drivers for application languages.

Cassandra writes all data first to a log to ensure that the data is durable in the event of failures. It then writes the data to in-memory structures that act as a cache, and when the cache gets full, flushes the data to disk into data files.

Cassandra automatically partitions all data and replicates it however many times you need. It uses various repair processes to ensure that the data across the nodes becomes consistent over time.

Periodically, Cassandra compacts the tables stored in the data files on disk, discarding any obsolete data it finds.

Cassandra randomly assigns client read and write requests to the cluster's nodes, with the initial node to which the client connects serving as the coordinator for that operation.

Understanding Cassandra-Specific Concepts

To master Cassandra administration, you should learn several Cassandra-specific structures and concepts, such as nodes and rings, commit logs, memtables, SSTables, keyspaces, index files, and bloom filters.

It's a good idea to start with a quick description of the essential high-level Cassandra structures.

- *Node*: A node is a server where a Cassandra instance runs. This is where Cassandra stores your data. You can run Cassandra on just a single node when starting out with the database, but all real-life Cassandra environments have a bunch of nodes, since Cassandra is designed for big data processing.

- *Datacenter*: The *datacenter* is a somewhat of a misleading term, as it has nothing to do with the term *data center* as you use it in everyday parlance, which refers to a physical or virtual data center of an organization.

  A Cassandra datacenter is a set of nodes that you configure as a group for replication proposes. Thus, you can view a datacenter as a synonym for a *replication group*. You use a datacenter to group a set of nodes that have the same replication level. A datacenter can be physical or logical. Typically, clients write to a local datacenter, which asynchronously replicates the data to other datacenters.

- *Cluster*: A Cassandra cluster is a set of one or more datacenters, and it can span physical locations.

Inside a Cassandra database, you need to understand the following key structures.

- *Commit Log*: Cassandra writes the data in memory to a commit log for durability. Since the database writes every write first to the commit log, it serves as a secure recovery mechanism and ensures that no writes are lost due to server or other failures.

- *Memtables*: After writing new data to the commit log, Cassandra writes that data to the memtables, which are data structures that live only in memory.

- *SSTable*: A sorted string table (SSTable) is an immutable data file to which Cassandra writes a memtable. Cassandra flushes all the data in the memtables to the SSTables once the memtables reach a threshold value.

- *CQL Table*: A CQL table is a set of ordered columns fetched by table row. Each table consists of a set of columns and a primary key, and it's the CQL tables that you query to retrieve data from a Cassandra database.

> **Tip**   It is not a good idea to use too many keyspaces or tables. Both entities have a JVM memory overhead. Each table uses approximately 1MB of memory, and an useful rule of thumb is to use no more than 1,000 tables per cluster, ideally just 500.

Cassandra uses several unique concepts and associated terminology. I think it's a good idea to get an overview of the concepts and the terminology before you plunge into the internals of the database in the following chapters.

Many of the terms and concepts I explain in the following sections are related to the management of a distributed database; there are various entities and strategies that Cassandra uses to manage and distribute data among multiple nodes, and how the nodes maintain awareness of the other nodes in the cluster.

Peers and Coordinators

All nodes in a Cassandra cluster are *peers*. A client's request can go randomly to any of the nodes in a cluster. When a client connects to one of the nodes in a cluster and issues a read or write request, that node will act as the coordinator for that operation.

The *coordinator* is a proxy between the client and the nodes (replicas) that own the data that the client's application requests. The coordinator node for a read or write operation determines to which node it should assign the request. It considers the partitioner you've configured and the replica's placement strategy.

Keyspaces

A *keyspace* is a collection of related tables and is analogous to the logical concept of a schema in an Oracle database, or a database when you're using MySQL, or the SQL Server database. The recommended practice is to use a single keyspace for each application.

Replication and the Replica Placement Strategy

Cassandra stores data on multiple nodes for reliability and fault tolerance. A *replica* is a copy of a chunk of data, and you set the replica placement strategy per keyspace when you create the keyspace.

Cassandra offers several replica placement strategies, such as the NetworkTopologyStrategy, which is ideal when you deploy your cluster across multiple data centers.

The number of replicas in each datacenter depends on the need to reduce cross-datacenter latency by promoting local reads and the need to provide protection against node failures.

If a node is down, the data that Cassandra failed to write to that node is replayed later, via the use of what are called *hinted handoffs,* which are hints that enable the writing of the missed data to the failed node when it comes back.

Tunable Consistency

As explained earlier in this chapter, Cassandra chooses availability over consistency. However, the degree of consistency is tunable by you. You can select the degree of consistency on a per-query basis when you're reading or writing data.

Tunable consistency means you pick how many replicas a read or write query should hear back from before Cassandra deems that read/write request as successful. When a client issues a read request, the consistency level determines how many replicas Cassandra must hear from before it returns the data to the client. As for a write request, it is the number of replicas that must confirm that they have successfully written the data to disk before Cassandra informs the client that the write was successful.

Obviously, a higher consistency level means slower reads and writes since Cassandra needs to wait for acknowledgement from more nodes. A higher level of consistency could also mean lower availability because you need more nodes to be available to satisfy the higher degree of consistency.

Gossip

Gossip refers to a communication protocol that a cluster's nodes use to discover other nodes and inform them of their own location and state, as well as that of other nodes in the cluster. The purpose of gossip is to let all nodes quickly get up-to-date information about the state of the rest of the nodes in the cluster.

It is from gossip data (state and history) that a node learns about other nodes that are down or have come back up after a failure.

Partitioner

Cassandra uses a function called a *partitioner* to determine which nodes receive the first copy of a chunk of data and how it should distribute the rest of the replicas across the other nodes.

The partitioner uses token values that it derives from the primary key of a row to make its determination of which nodes get the replicas of a specific row. Cassandra offers various partitioners, with the default partitioner being the Murmur3Partitioner.

Snitch

To place the replicas of data on the nodes, Cassandra uses a topology. A *snitch* defines the topology; that is, it groups machines into datacenters and racks.

Cassandra offers multiple snitches. The default *SimpleSnitch* is for simple installations that don't use a complex topology that uses datacenters and racks. For production deployments, Cassandra recommends that you use the *GossipingPropertyFileSnitch*. This snitch defines datacenters and racks.

Compaction

Compaction is how Cassandra periodically consolidates data stored in tables in data files on disk. During the compaction process, Cassandra collects all the versions of a row, and from them assembles the most up-to-date versions of that row. It then writes the new row versions to a new SSTable and leaves the old versions along with other rows that are ready for deletion in the old SSTables. As soon as all pending reads are completed, the database deletes the old versions, using markers called tombstones, which indicate that the data is deleted data.

Bloom Filters

Bloom filters are algorithms for testing whether an element is a member of a set. Cassandra employs bloom filters to quickly determine whether a table (SSTable) has the data for a row. It uses the clusters for index scans but not for range scans.

Node Repairs

Since Cassandra is a distributed database, over time data in a replica can become inconsistent with other replicas stored in other nodes.

Node repair is a regular maintenance task that corrects the inconsistencies among the replicas and works towards establishing eventual consistency; that is, it ensures that over time all nodes have the same, most recently updated data.

There are several types of repair processes, such as a hinted handoff, where Cassandra preserves the data to be written to a node as a set of hints, when that node can't receive any writes. When the node comes back up, the database uses the stored set of hints to perform a repair so that the node can add the writes that it missed while it was unavailable.

How Cassandra Stores Its Data

At a high level, Cassandra stores its data in the form of a hash *ring*, with each node in the cluster owning a range of hashes. Cassandra partitions the data around the ring, and the location of data on the hash ring is based on the partition key.

All nodes are equal (no master) and continuously communicate with each other, exchanging their state. All nodes participate in all reads and writes.

Data is replicated to multiple nodes, and you can configure the number of copies, with the default being three replicas.

An Overview of Cassandra's Data Model

Cassandra's data model sharply differs from that of a traditional relational database. Table 1-1 shows the key differences between the two data models.

Table 1-1. *How Cassandra's Data Model Differs From That of an RDBMS*

RDBMS	Cassandra
Structured data	Unstructured data
Fixed schema	Flexible schema
A row is a single record.	A row is a unit of replication.
Columns represent a relation's attributes.	Columns are a unit of storage.
Uses foreign keys and joins	Uses collections to represent relationships
A table is an array of arrays.	A table is a list of nested (rows by columns) key-value pairs.
The database is the outermost data container.	The keyspace is the outermost container for data.

DataStax and Cassandra

DataStax, Inc. is a company that offers a commercial version of the Apache Cassandra database called DataStax Enterprise (DSE). DataStax 1.0 was released in late 2011. DataStax is the leading evangelizer for the Cassandra database, with several employees contributing to the open-source Cassandra project.

Besides the commercial DSE, DataStax also offers a free distribution of Apache Cassandra called the Community Edition.

In the following sections, I briefly review DSE and the DataStax OpsCenter, a management tool for configuring and managing Cassandra.

DataStax Enterprise

DataStax Enterprise provides a comprehensive and simple data management layer with an always-on architecture built on top of Apache Cassandra. DSE offers advanced functionality to create sophisticated cloud applications.

DSE integrates several powerful features such as indexing, search, analytics, and graph functionality with the open source Apache Cassandra database. Here is a brief description of the features that DSE offers on top of the underlying Cassandra database.

- *Advanced Indexing and Search*: DSE Search contains an advanced indexing engine that contains powerful search capabilities (DSE integrates Apache Solr, a well-known search tool) to help find data by running complex queries. The indexing engine supports fast, real-time aggregations, filtering, faceting, as well as sub-string, and fuzzy and full text search.

- *Powerful Integrated Analytics*: DSE Analytic integrates the powerful Apache Spark large-scale data processing engine to enable developers to build pipelines that interact with data flowing through both streaming and batch workloads.

- *Graph Database*: DSE Graph is a scalable, real-time graph database that supports deep analytical queries. Graph databases are designed for analyzing large data sets with many complex relationships by helping you discover and make sense of those relationships.

Organizations trying to derive the most value from their data have different requirements, each solved by a different solution, such as advanced analytics, real-time indexing, search, etc. DSE offers all the capabilities required by modern cloud applications in an integrated fashion, through one tool. Integration of search, analytics, and graph database capabilities means that these components derive all the benefits of the Apache Cassandra database, such as high availability, linear scalability, and predictable, low-latency response times.

The DataStax Enterprise OpsCenter is a web-based management and monitoring component of DSE. OpsCenter makes it easy to monitor, tune, backup, and recover the Cassandra database. You can perform most of your database management tasks, such as access and authorization, through OpsCenter. OpsCenter provides a centralized dashboard to monitor Cassandra clusters and helps proactively identify issues before they affect your production workloads. You can easily integrate OpsCenter into management tools such as Graphite.

DataStax Development Tools

DataStax offers several tools to help you create Cassandra-based, large-scale distributed applications. The following is a brief description of the three key development tools offered by DataStax: the DataStax DevCenter, the DataStax Studio, and database drivers.

- *DataStax Enterprise OpsCenter*: Web-based visual management and monitoring solution for DataStax Enterprise (DSE). This tool helps you manage database schemas, develop queries, and tune Cassandra performance.

- *DataStax Studio*: The DataStax Studio is an interactive tool that uses DSE Graph to explore and visualize large data sets.

- *Database Drivers*: DataStax offers database drivers that support languages such as C#, Java, Node.js, ODBC, Python, PHP, and Ruby.

Summary

Cassandra is a powerful NoSQL database that significantly differs from a relational database. This chapter explained the ways in which Cassandra differs from a relational database, and how it solves problems that relational databases don't handle well.

Cassandra is designed for specific use cases, and if your needs are different, it may be an inappropriate solution for you.

Understanding the ACID requirements and the CAP theorem helps you realize the significance of Cassandra's eventual consistency principle.

Although Cassandra shares some optimization strategies with relational databases, it also makes use of unique strategies such as bloom filters to optimize database performance.

The chapter reviewed the essentials of what a Cassandra database administrator ought to know.

Understanding Cassandra specific concepts, such a snitches, gossip, and replication strategies, is critical to your success as a Cassandra administrator.

CHAPTER 2

Installing Cassandra and Getting Started with CQL Shell

In this chapter, you'll learn how to install Cassandra and create a single-node Cassandra "cluster." Once you get your feet wet, it's an easy transition to creating and configuring a multi-node Cassandra cluster, which is the topic I discuss in Chapter 3.

Once you create a Cassandra cluster, you learn how to work with cqlsh, the command line interface to the CQL shell. By the time you complete this chapter, you'll know how to start and stop your cluster, how to create keyspaces and tables, and how to insert and query data.

Installing Apache Cassandra

Cassandra is easy to set up and get going with. In this section, I show how to install Cassandra on Linux, specifically on an Ubuntu 16.04 LTS server.

Once you learn how to install Cassandra, I show how to create a simple one-node Cassandra cluster on a single machine. In Chapter 3, I show how to create a Cassandra cluster with multiple nodes.

Planning the Installation

I use an Ubuntu 16.04 LTS server running on VMware to show the installations. The server has 4GB of RAM and 20GB of storage.

© Sam R. Alapati 2018
S. R. Alapati, *Expert Apache Cassandra Administration*, https://doi.org/10.1007/978-1-4842-3126-5_2

Before you start with the installation of Cassandra, you need to create a user and group that will manage Cassandra on your server. Name the user *cassandra* and the group *cassandra* as well. Here are the steps to create the user and the group:

```
$ sudo groupadd -r cassandra
$ sudo useradd -r -m cassandra -g cassandra -G users
```

Prerequisites for Installing Cassandra

There are two basic prerequisites that you need before you can install Cassandra: you must install the correct version of Java, and you need the latest version of Python.

The following two sections show how to install Java and Python, should it be necessary to do so.

Java

You need the latest 64-bit version of Java 8, and either the Oracle Java Standard Edition or OpenJDK 8 will do (I use the Oracle JRE). You can check the version of the installed Java software thus:

```
$ java -version
java version "1.8.0_101"
Java(TM) SE Runtime Environment (build 1.8.0_101-b13)
Java HotSpot(TM) 64-Bit Server VM (build 25.101-b13, mixed mode)
$
```

If the output doesn't show something similar to "*java version "1.8.0_101"* on the first line (there are three lines in the output), you need to install Java. Follow these steps to install the Oracle JDK.

1. Download the Oracle JDK installer from Oracle Java SE Downloads, after accepting a license agreement. You can download the JDK by going here:

   www.oracle.com/technetwork/java/javase/downloads/
   jdk8-downloads-2133151.html

2. Make a directory for the JDK.

   ```
   $ sudo mkdir -p /usr/lib/jvm
   ```

3. Install the JDK after unpacking the tarball.

```
$ sudo tar zxf jdk-8u65-linux-x64.tr.gz -C /usr/lib/jvm
```

You can find the JDK files in the directory named /usr/lbi/jvm/ jdk-8u-version.

4. You need to let the OS know about the new Java versions with the alternatives command.

```
$ sudo update-alternatives -install "/usr/bin/java" "java"
"/usr/lib/jvm/jdk1.8.0_101/bin/java"
```

5. Run the alternatives command again, this time to make the new JDK the default Java release.

```
$ sudo update-alternatives -config java
```

This command makes sure that if you have multiple Java installations, the default version is switched to the release you just installed.

6. Confirm that the install was correct.

```
$ java -version
java version "1.8.0_101"
Java(TM) SE Runtime Environment (build 1.8.0_101-b13)
Java HotSpot(TM) 64-Bit Server VM (build 25.101-b13,
mixed mode)
$
```

Python

While you need Java for running Cassandra itself, you need Python for running cqlsh, the command line interface to the Cassandra database. To be specific, you need the latest version of Python 2.7.

You can check the current release of Python thus:

```
$ python
Python 2.7.12 (default, Nov 19 2016, 06:48:10)
[GCC 5.4.0 20160609] on linux2
Type "help", "copyright", "credits" or "license" for more information.
>>>
```

In my case, the Ubuntu server came with Python. If you need to install Python, you can do so by following these steps.

1. Install the required packages.

   ```
   $ sudo apt-get install build-essential checkinstall
   $ sudo apt-get install libreadline-gplv2-dev libncursesw5-
   dev libssl-dev libsqlite3-dev tk-dev libgdbm-dev libc6-dev
   libbz2-dev
   ```

2. Download the Python binaries.

   ```
   $ cd /usr/src
   $ wget https://www.python.org/ftp/python/2.7.12/
   Python-2.7.12.tgz
   ```

3. Extract the packages.

   ```
   $ tar xzf Python-2.7.12.tgz
   ```

4. Compile the Python source.

   ```
   $ cd Python-2.7.12
   $ sudo ./configure
   $ sudo make altinstall
   ```

5. Check the Python version.

   ```
   $ python2.7 -V
   Python 2.7.13
   $
   ```

Installing Cassandra

You can install Cassandra from a source binary tarball that you download, install, and run, or use yum or apt-get to install the Cassandra package as a service (Linux only). In this section, I show both ways of installing Cassandra. You can also install using a less common third method, which is to download the source code and compile it.

Building from Source

You can build Cassandra from the source by using Apache Ant. When you build from source, you need to use a Java 8 JDK instead of just the JRE. Here are the steps to build from source.

1. Download Ant if you don't have it already, from

    ```
    http://ant.apache.org
    ```

2. Get a read-only trunk version of the Cassandra source.

    ```
    $ git clone git://git.apache.org/cassandra.git
    ```

3. Move to the root directory of the source download and run ant.

    ```
    $ ant
    ```

Ant looks for a `build.xml` file fin the current directory and executes the default build target. Ant builds the source files and executes tests, and if everything goes well, you'll see a BUILD SUCCESSFUL message.

Installing from Debian Packages

Installing from Debian packages is just as simple as installing from a source binary tarball, as explained in the following steps.

1. Add Cassandra's Apache repository to `/etc/apt/sources.list.d/cassandra.sources.list`, as shown here:

    ```
    $ echo "deb http://www.apache.org/dist/cassandra/
    debian 311x main" | sudo tee -a /etc/apt/sources.list.d/
    cassandra.sources.list
    ```

2. Add the Apache Cassandra repository keys.

    ```
    $ curl https://www.apache.org/dist/cassandra/KEYS |
    sudo apt-key add -
    $ sudo apt-key adv --keyserver pool.sks-keyservers.net
    --recv-key A278B781FE4B2BDA
    ```

3. Update the package index.

   ```
   $ sudo apt-get update
   ```

4. Now you're ready to install Cassandra.

   ```
   $ sudo apt-get install cassandra
   ```

5. You can start, stop, and check the status of the Apache Cassandra service on this server by running the following command:

```
$ systemctl start cassandra.service$
$ sudo systemctl stop cassandra.service
$ sudo systemctl status cassandra.service
[sudo] password for samalapati:
• cassandra.service - LSB: distributed storage system for structured data
  Loaded: loaded (/etc/init.d/cassandra; bad; vendor preset: enabled)
  Active: active (running) since Fri 2017-09-29 08:07:57 PDT; 42min ago
    Docs: man:systemd-sysv-generator(8)
   Tasks: 57
  Memory: 2.2G
     CPU: 40.810s
  CGroup: /system.slice/cassandra.service
          └─5432 java -Xloggc:/var/log/cassandra/gc.log
            -ea -XX:+UseThreadPrior

Sep 29 08:07:57 ubuntu systemd[1]: Starting LSB: distributed storage
system for
Sep 29 08:07:57 ubuntu systemd[1]: Started LSB: distributed storage
system for s
Sep 29 08:08:11 ubuntu systemd[1]: Started LSB: distributed storage
system for s
```

6. You can enable the Cassandra service on system boot by doing the following:

   ```
   $ sudo systemctl enable cassandra.service
   ```

7. Since the Debian packages start up the Cassandra service automatically after the installation, you must stop the service and clear the data if you're planning a real-life cluster rather than a test service. By doing this, you remove the default cluster name (Test Cluster) from Cassandra's system table. Here's how to clear the data:

```
$ sudo systemctl service cassandra.service stop
$ sudo rm -rf /var/lib/cassandra/data/system/*
```

Building from the Source Binary Tarball

The easiest way to get going with installing Cassandra is to download and install the binary tarball from Cassandra's website:

http://cassandra.apache.org/download/

You can also download the latest tarball by using a command such as curl:

```
$ curl -OL http://www.apache.org/dist/cassandra/3.9.0/apache-cassandra-
3.9.0-bin.tar.gz
```

Follow these steps to download and install Cassandra from binary tarball files.

1. Download the latest Cassandra release, which is 3.9 (release 2016-09-29).

2. Extract the installation files from the tarball.

    ```
    $ sudo tar -xzvf apache-cassandra-3.9-bin.tar.gz
    ```

    The tar command will extract the files into the apache-cassadnra-3.9 directory.

3. Move the extracted directory to the /usr/share library.

    ```
    $ sudo mv apache-cassandra-3.9 /usr/share
    ```

4. Create a symbolic link to the directory as shown here:

    ```
    $ sudo ln -s /usr/share/apache-cassandra-3.9 /usr/share/
    cassandra
    ```

Creating the symbolic link is a good idea for down the road. The symbolic link allows you to keep multiple versions of Cassandra in the same installation. All you have to do to switch between the versions is to simply change the location the link points to.

That's it; you're done! Unlike installing a relational database and configuring it, it's easy to install Cassandra and get going with it. With no configuration at all, it's possible to start working with Cassandra out of the box. Of course, you'll need to configure and tune Cassandra to get the most out of it, but it's quite easy for a novice to get started with the database.

In the next section, I show how to configure Cassandra by entering and modifying the configuration properties in `cassandra.yaml`, the main Cassandra configuration file. Later, I show you how to start and stop Cassandra.

Creating the Necessary Directories

You need to create a basic set of three directories so Cassandra knows where to store the table and other data. Create the following three data directories as the user *cassandra*:

```
$ mkdir /usr/share/cassandra/commitlog
$ mkdir /usr/share/cassandra/data
$ mkdir /usr/share/cassandra/saved_caches
```

Cassandra will create these three directories regardless of whether you create them; the reason you want to do this yourself is so the directories are located in a location of your choice, instead of in the default directories.

Configuring Cassandra

The `cassandra.yaml` file is the key configuration file for setting the configuration properties for a Cassandra cluster, such as the following:

- Tuning and resource utilization parameters
- Caching parameters
- Client connections
- Backups
- Security

Cassandra offers a `cassandra.yaml` template, located in the $CASSANDRA_HOME/ conf directory. When starting out with Cassandra, you can just edit a handful of the configuration properties to get going. Later, once you understand Cassandra's architecture and the key concepts well, it'll be time to tinker with the configuration properties that affect performance and many other aspects of Cassandra, which I explain in the relevant chapters in this book.

Location of the cassandra.yaml File

By default, the `cassandra.yaml` file is located in the following locations:

- `/etc/cassandra`   `/* for Cassandra package installations`

- `<install_location>/conf` `/* for Cassandra tarball installations`

Minimal Configuration Properties You Must Set

Although you can configure hundreds of initialization parameters in the `cassandra. yaml` file, to start with you need to configure just a handful of properties:

- `cluster_name`

- `listen_adresss`

- `listen_interface`

I explain these minimal properties in the following sections. You can save the default `cassandra.yaml` with a different name and create a new `cassandra.yaml` file with just these properties to start your first single-node Cassandra cluster. You specify each configuration property by placing a colon after the property.

The cluster_name Parameter

The `cluster_name` parameter lets you name the Cassandra cluster. The default value is Test Cluster. Obviously, you must specify the same `cluster_name` value for all the nodes in a multi-node cluster.

The main purpose of the `cluster_name` parameter is to prevent nodes belonging to one logical cluster from joining other clusters.

The listen_adresss Property

The listen_address property determines the IP address (or hostname) that Cassandra binds to when connecting to other nodes. The default value is localhost. If you are running a single-node cluster, you can just use the default setting for this parameter.

Caution   Don't set the value of 0.0.0.0 for the listen_address parameter.

The listen_interface Parameter

The listen_interface parameter specifies the interface that Cassandra binds to when connecting to other nodes and must correspond to a single address.

If you set the listen_address property, you can omit the listen_interface parameter and vice versa.

Setting the Data File Directory Locations

You specify the directories for storing the Cassandra data files as well its cache directories in the cassandra.yaml file.

There are two main directories you must configure to specify where Cassandra stores its data; the first location is for storing the SSTables, and the second location is for storing the commit log. In addition, there's a third directory for storing cache data, called the saved_caches directory.

Although Cassandra will create these directories for you in their default locations, you can also set a custom location for the following three Cassandra directories, as shown here:

- data_file_directories: /usr/share/cassandra/data

- commitlog_directory: /usr/share/cassandra/commitlog

- saved_caches_directory: /usr/share/cassandra/saved_caches

By default, Cassandra stores its data in the following locations:

- /var/lib/cassandra   /* for Package installations

- <install_location>/data/data   /* for Tarball installations

44

If you don't set this parameter, the default directory is $CASSANDRA_HOME/data/data.

Location for Storing the Cache Directory

Cassandra needs a directory to store its key and row caches. The default location for this is

- `/var/lib/cassandra/saved_caches    /* for Package installations`

- `<install_location>/data/saved_caches /* for Tarball installations`

Setting the Location for Storing the Commit Log

You should assign a different directory, ideally on a different spindle, for storing the commit log.

By default, Cassandra stores the commit log in the /var/lib/cassandra/commitlog directory. If you don't set it, the default directory will be the following:

$CASSANDRA_HOME/data/commitlog

Configuring the Firewall

Before you start up the Cassandra instance, make sure you can access the Cassandra services from outside the server on which Cassandra is running, by opening these ports:

- 7000

- 7199

- 9042

- 9160

Exploring the CQL Shell

The Cassandra Query Language (CQL) is the primary means of communication with the Cassandra database. The simplest way to interact with Cassandra is to use the CQL shell, cqlsh.

You create keyspaces, tables, and read and write data via cqlsh. In the following sections, I show you how to get started with cqlsh.

Starting the CQL Shell

You start the CQL shell with the `cqlsh` command, as shown here:

```
$ cqlsh
Connected to Test Cluster at 127.0.0.1:9042.
[cqlsh 5.0.1 | Cassandra 3.7 | CQL spec 3.4.2 | Native protocol v4]
Use HELP for help.
cqlsh>
```

You terminate the CQL shell by typing `exit` at the cqlsh prompt:

```
cqlsh:mykeyspace1> exit;
$
```

Time Zones in cqlsh

Cassandra displays timestamps with an UTC timezone by default. You must install the *pytz* library to display timestamps with a different timezone.

Getting Help in the CQL Shell

The HELP facility in the CQL shell is cool. When you type HELP at the cqlsh command line, you see all the options for HELP, as is the case with any other command line utility.

```
cqlsh> help
Documented shell commands:
=============================
CAPTURE  CLS          COPY  DESCRIBE  EXPAND  LOGIN   SERIAL  SOURCE   UNICODE
CLEAR    CONSISTENCY  DESC  EXIT      HELP    PAGING  SHOW    TRACING
```

```
CQL help topics:
=================
AGGREGATES                  CREATE_KEYSPACE             DROP_TRIGGER          TEXT
ALTER_KEYSPACE              CREATE_MATERIALIZED_VIEW    DROP_TYPE             TIME
ALTER_MATERIALIZED_VIEW     CREATE_ROLE                 DROP_USER        TIMESTAMP
ALTER_TABLE                 CREATE_
                            TABLE                       FUNCTIONS         TRUNCATE
ALTER_TYPE                  CREATE_TRIGGER              GRANT                TYPES
ALTER_USER                  CREATE_TYPE                 INSERT              UPDATE
APPLY                       CREATE_USER                 INSERT_JSON            USE
ASCII                       DATE                        INT                   UUID
BATCH                       DELETE                      JSON
BEGIN                       DROP_AGGREGATE              KEYWORDS
BLOB                        DROP_COLUMNFAMILY           LIST_PERMISSIONS
BOOLEAN                     DROP_FUNCTION               LIST_ROLES
COUNTER                     DROP_INDEX                  LIST_USERS
CREATE_AGGREGATE            DROP_KEYSPACE               PERMISSIONS
CREATE_COLUMNFAMILY         DROP_MATERIALIZED_VIEW      REVOKE
CREATE_FUNCTION             DROP_ROLE                   SELECT
CREATE_INDEX                DROP_TABLE                  SELECT_JSON
cqlsh>
```

In addition, when you pass an option with the HELP command, such as HELP CREATE_TABLE, Cassandra brings up a nice screen showing the syntax for the command, as shown in Figure 2-1.

CREATE TABLE

Syntax:

```
<create-table-stmt> ::= CREATE ( TABLE | COLUMNFAMILY ) ( IF NOT EXISTS )? <tablename>
                        '(' <column-definition> ( ',' <column-definition> )* ')'
                        ( WITH <option> ( AND <option>)* )?

<column-definition> ::= <identifier> <type> ( STATIC )? ( PRIMARY KEY )?
                      | PRIMARY KEY '(' <partition-key> ( ',' <identifier> )* ')'

<partition-key> ::= <identifier>
                  | '(' <identifier> (',' <identifier> )* ')'

<option> ::= <property>
           | COMPACT STORAGE
           | CLUSTERING ORDER
```

Figure 2-1. *The HELP command for the CREATE TABLE statement*

Note   cqlsh is guaranteed to work well (that is, be fully compatible with) just the Cassandra release cqlsh is belongs to. It might work well with both older and newer releases of Cassandra, but don't count on it!

Command Line CQL Shell Options

You'll be using cqlsh to perform many tasks, so you should get a good grip on the basic CQL shell command line options. The cqlsh utility comes with several useful command line options, also called *documented shell commands*. I'll show how to use the most useful CQL shell options here.

The capture Command

The CAPTURE command will capture the output of a command and add it to a text file, as shown in the following example:

```
cqlsh> capture '/home/test/Cassandra/output/testfile'
```

Once you issue the CAPTURE command, the shell will capture the output of all subsequent CQL commands that you issue in that session, until you turn off the output capturing with the capture off command:

```
cqlsh> capture off;
```

The copy Command

The COPY command is great for capturing the data from a Cassandra table to a text file. The following is an example that shows how to capture the contents of a table named employees to the file myfile:

```
cqlsh> copy employee (emp_id, emp_cicty, emp_name, emp_sal) to 'myfile';
```

The describe Command

The DESCRIBE command will describe various entities in a Cassandra cluster. In the following example, the describe cluster command describes the Cassandra cluster and its top-level objects:

```
cqlsh> describe cluster;
Cluster: Test Cluster
Partitioner: Murmur3Partitioner
cqlsh>
```

You can get a description of several other entities with the DESCRIBE command:

- Describe types: Lists all user-defined data types

- Describe type: Describes a user-defined data type

- Describe tables: Lists all the tables in a keyspace

- Describe table: Describes a table

- Describe keyspaces: Lists all keyspaces in a cluster

Note   The DESC command works the same as the DESCRIBE command.

Here's an example showing how to run the DESCRIBE KEYSPACES command:

```
cqlsh> describe keyspaces;
test1           system_auth   system_distributed   testdata
system_schema   system        system_traces
cqlsh>
```

Let's say I have a table named quarters. When I run the describe table quarters command, I get a bunch of output, although I only used a couple of lines to create the table.

```
cqlsh> use testdata;
cqlsh:testdata> describe table quarters;

CREATE TABLE testdata.quarters (
    id int PRIMARY KEY,
    name text
) WITH bloom_filter_fp_chance = 0.01
    AND caching = {'keys': 'ALL', 'rows_per_partition': 'NONE'}
    AND comment = ''
    AND compaction = {'class': 'org.apache.cassandra.db.compaction.
    SizeTieredCompactionStrategy', 'max_threshold': '32', 'min_threshold': '4'}
    AND compression = {'chunk_length_in_kb': '64', 'class': 'org.apache.
    cassandra.io.compress.LZ4Compressor'}
    AND crc_check_chance = 1.0
    AND dclocal_read_repair_chance = 0.1
    AND default_time_to_live = 0
    AND gc_grace_seconds = 864000
    AND max_index_interval = 2048
    AND memtable_flush_period_in_ms = 0
    AND min_index_interval = 128
    AND read_repair_chance = 0.0
    AND speculative_retry = '99PERCENTILE';
cqlsh:testdata>
```

If you are coming from a relational database such as Oracle, you can recognize just one of these table options, PRIMARY KEY, which you must specify when you create a table. The rest are all default values. This is what makes Cassandra so much fun, since there is a lot of new and interesting stuff to learn and use. You'll learn about all these table options in the appropriate chapters.

The expand Command

The expand command shows the contents of a table's rows vertically, making it convenient to read long rows of data. Instead of scrolling to the right, as in the case of the default horizontal format, you scroll down to view more of a row.

Let's say your query results in the following output:

```
cqlsh:mykeyspace1> select * from employee;
 emp_id | emp_city    | emp_name | emp_phone    | emp_sal
--------+-------------+----------+--------------+---------
      1 | San Antonio |     juan | 39874622562  |   90000
      2 |     Houston |      jim | 87209887521  |  100000
      3 |      Austin |      sam | 87361598012  |   50000
(3 rows)
cqlsh:mykeyspace1>
```

Now you would like to get a more detailed output. You can do this by first issuing the expand on command in the cqlsh session, and following it up with your original command.

```
cqlsh:mykeyspace1>
expand on;
Now Expanded output is enabled
cqlsh:mykeyspace1> select * from employee;
@ Row 1
-----------+-------------
 emp_id    | 1
 emp_city  | San Antonio
 emp_name  | juan
 emp_phone | 39874622562
 emp_sal   | 90000

@ Row 2
-----------+-------------
 emp_id    | 2
 emp_city  | Houston
 emp_name  | jim
 emp_phone | 87209887521
 emp_sal   | 100000
```

```
@ Row 3
-----------+-------------
 emp_id    | 3
 emp_city  | Austin
 emp_name  | sam
 emp_phone | 87361598012
 emp_sal   | 50000

(3 rows)
cqlsh:mykeyspace1>
```

Be sure to turn expanded output off once you're done with it:

```
cqlsh:mykeyspace1> expand off;
Disabled Expanded output.
cqlsh:mykeyspace1>
```

The tracing Command

The tracing command lets you enable and disable tracing of transactions running in the database. You use tracing to diagnose performance issues. The system_traces keyspace captures the information about the Cassandra internal operations. The table named session in this keyspace captures the query results and high-level details. Cassandra captures the detailed information for all operations it performs in the system_traces. events table. I explain tracing in detail in Chapter 11.

Cassandra Installation Directories

When you install Cassandra, the installer creates a bunch of directories and it's a good idea to learn what those directories are and what's in them.

The location of the directories depends on whether you install Cassandra through a tarball or a package. In general, a tarball installation creates all the directories under the installation directory, and a package install will store its directories under the /etc and / var directories.

In my case, I installed Cassandra using a tarball, so I'll show the directory structure in my installation, which is as follows:

```
$ cd $CASSANDRA_HOME
$ ls
bin             conf  interface  lib           NEWS.txt    pylib
CHANGES.txt  doc   javadoc    LICENSE.txt   NOTICE.txt  tools
$
```

The following sections describe the key Cassandra directories and the important files under those directories.

The bin Directory

The bin directory contains various utilities and start scripts, as shown here:

```
$ ls
cassandra          cqlsh.py         sstableloader       sstableutil.bat
cassandra.bat      debug-cql        sstableloader.bat   sstableverify
cassandra.in.bat   debug-cql.bat    sstablescrub        sstableverify.bat
cassandra.in.sh    nano.save        sstablescrub.bat    stop-server
cassandra.ps1      nodetool         sstableupgrade      stop-server.bat
cqlsh              nodetool.bat     sstableupgrade.bat  stop-server.ps1
cqlsh.bat          source-conf.ps1  sstableutil
$
```

The bin directory is where you'll find the following key utilities:

- cassandra: This utility helps you start a Cassandra instance, as I explain in the next section.

- cqlsh: The cqlsh utility starts up the CQL shell that enables you to write CQL queries to talk to Cassandra.

- nodetool: The nodetool utility is a Cassandra administrator's best buddy. It lets you perform numerous administrative tasks such as checking the status of the cluster, decommissioning nodes, adding nodes, etc. Throughout this book, I use nodetool to perform numerous administrative tasks.

- sstableloader: This tool enables you to load data into SSTables, as I explain in Chapter 9.

The tools Directory

The `tools` directory contains useful Cassandra tools for tasks such as stress testing and for managing SSTables, as shown here:

- cassandra-stress: Cassandra's load testing tool

- sstabledump: Utility that dumps the contents of an SSTable in JSON format

- sstablesplit: Tool to split SSTables into multiple tables

- sstablemetadata: Prints metadata about an SSTable.

The lib Directory

The `lib` directory contains all the external libraries that Cassandra may need during its execution, such as JSON serialization libraries and the Apache commons libraries.

The conf Directory

The `conf` directory contains everything you need to configure a Cassandra cluster. It contains the configuration files that enable you to set the runtime properties for Cassandra nodes as well as for the cluster racks. This directory also contains the files where you set the environment for the database. Here's a quick description of the key files in this directory:

- `cassandra.yaml`: This is the main Cassandra configuration file.

- `cassandra-env.sh`: This is the file where you configure the Linux settings for Java, the JVM, and the JMX.

- `cassandra-rackdc.properties`: This file defines the default datacenter and rack used by various snitches such as GossipingPropertyFileSnitch, Ec2Snitch, Ec2MultiRegionSnitch, and GoogleCloudSnitch.

- `Cassandra-topology.properties`: This file defines the default datacenter and rack for the `PropertyFileSnitch`.

- `jvm.options`: Allows you to set the options that Cassandra will use to start the JVM.

- `commitlog_archiving.properties`: Enables the configuration of the commit log.

- `metrics-reporter-config-sample.yaml`: An example file that shows how to configure Cassandra metrics.

- `logback.xml`: Logback configuration file that helps you configure Cassandra's logging settings.

The Javadoc Directory

The `javadoc` directory contains a documentation website that was generated using the JavaDoc tool. This is not a complete documentation, but rather the comments that are stored in the Java code. To read the JavaDoc, open the `javadoc/index.html` file in your browser.

Starting and Stopping Cassandra

You start and stop Cassandra by issuing the `cassandra` command. The cassandra utility, along with other key Cassandra administrative tools, is located in the `$CASSANDRA_HOME/bin` directory.

Starting Cassandra

You can start up Cassandra with the `cassandra` command. You can optionally add the following flags:

- `-f` starts Cassandra in the foreground (by default, the `cassandra` command starts the database in the background). Running in the foreground means the server will print all logs to the standard output and you can see them in your terminal window.

- Cassandra always writes the server logs to the system.log file, regardless of whether you run Cassandra in the background or in the foreground.

- -R starts Cassandra as the root user.

Here's an example of how to start a Cassandra instance (in the foreground, as the user root):

```
$ cassandra -f -R
...
INFO  20:33:47 Configuration location: file:/cassandra/apache-
cassandra-3.9/conf/cassandra.yaml
data_file_directories=[Ljava.lang.String;@175c2241; disk_access_mode=auto;
disk_failure_policy=stop; dynamic_snitch=true; dynamic_snitch_badness_
threshold=0.1; dynamic_snitch_reset_interval_in_ms=600000;
...
INFO  20:33:57 Initializing system_schema.keyspaces
INFO  20:33:59 Cassandra version: 3.9
INFO  20:34:00 Loading persisted ring state
INFO  20:34:00 Starting up server gossip
INFO  20:34:00 Updating topology for localhost/127.0.0.1

INFO  20:34:01 Node localhost/127.0.0.1 state jump to NORMAL

INFO  20:34:01 Starting listening for CQL clients on
localhost/127.0.0.1:9042 (unencrypted)...
```

I ran the cassandra command with the −f option, thus choosing to run the instance in the foreground. Therefore, Cassandra will keep printing all logging information in the terminal. If you start Cassandra in the default background mode, you can get the Linux command prompt back after starting the Cassandra instance by pressing Enter.

Let's review the abbreviated output I've presented here, as it teaches us a few valuable things:

- *Loading persisted ring state*: Loads the ring state.

- *Staring up server gossip*: Statements pertaining to gossip indicate that the server is initiating communications with the rest of the cluster's nodes.

- *Updating topology*: Updates the cluster topology by adding any new nodes that you've added to the cluster.

- *Node ... state jump to NORMAL*: This means that Cassandra has started up fine and is waiting for you to work with it through cqlsh or by other means.

Once the server starts, Cassandra continually keeps writing to the `system.log` file, updating it with information pertaining to internal database activities such as the flushing of memtables and the compaction of SSTables.

Checking Cassandra's Status

You can check the status of an instance with the `nodetool status` command. The nodetool utility works only when the Cassandra instance is running on a node. Nodetool is an immensely useful Cassandra tool, and it's a good idea to familiarize yourself with it.

Type `nodetool -help` to view all nodetool's commands that you can execute:

```
$ nodetool -help
usage: nodetool [(-h <host> | --host <host>)] [(-p <port> | --port <port>)]
        [(-pwf <passwordFilePath> | --password-file <passwordFilePath>)]
        [(-u <username> | --username <username>)]
        [(-pw <password> | --password <password>)] <command> [<args>]

The most commonly used nodetool commands are:
    assassinate                 Forcefully remove a dead node without
                                re-replicating any data.  Use as a last
                                resort if you cannot removenode
    bootstrap                   Monitor/manage node's bootstrap process
    cleanup                     Triggers the immediate cleanup of keys no
                                longer belonging to a node. By default,
                                clean all keyspaces
    clearsnapshot               Remove the snapshot with the given
                                name from the given keyspaces. If no
                                snapshotName is specified we will remove
                                all snapshots
    compact                     Force a (major) compaction on one or more
                                tables or user-defined compaction on given
                                SSTables
```

compactionhistory	Print history of compaction
compactionstats	Print statistics on compactions
decommission	Decommission the *node I am connecting to*
describecluster	Print the name, snitch, partitioner and schema version of a cluster

...

If you run nodetool when there's no Cassandra instance running, you'll get an error:

```
$ nodetool status
nodetool: Failed to connect to '127.0.0.1:7199' - ConnectException:
'Connection refused'.
$
```

The nodetool status Command

The nodetool status command lets you learn about the status of a Cassandra cluster. Since I have the Cassandra instance running now, I can use nodetool to check on it:

```
$ nodetool status
Datacenter: datacenter1
=======================
Status=Up/Down
|/ State=Normal/Leaving/Joining/Moving
--  Address          Load      Tokens      Owns (effective)  Host ID
                     Rack
UN  192.168.177.132  203.5 KiB  256           100.0%           b0ade950-
937a-457c-95eb-d3032897eeb1  rack1
$
```

Each node in the output is represented by its IP address. The very first column in nodetool's status report shows the status and the state of the Cassandra instance.

Status can take two values: Up or Down.

State can take one of the following four values:

- Normal

- Leaving

- Joining

- Moving

In my case, I see a UN, which means Up/Normal. The Owns columns shows the percentage of data owned by the node per data center times the replication factor of the data. If for example, a node owns 33% of the data, the Owns column shows 67%, if the replication factor is 2. Since right now there's just a single node in my cluster, it owns all (100%) of the data in the cluster. A bad data model will affect the distribution of data among the nodes of a cluster, and checking the percentage of data owned by each node is a good way to get to check if the data model is good.

Testing the Server with the nodetool info Command

The nodetool info command, which shows information such as uptime and load, helps you verify that the Cassandra instance is running properly, as shown here:

```
$ sudo nodetool info
ID                       : 99c43633-c691-4dee-b7af-35bc6e74dd67
Gossip active            : true
Thrift active            : false
Native Transport active: true
Load                     : 296.85 KiB
Generation No            : 1489357081
Uptime (seconds)         : 1714
Heap Memory (MB)         : 120.53 / 1014.00
Off Heap Memory (MB)     : 0.00
Data Center              : datacenter1
Rack                     : rack1
Exceptions               : 21
Key Cache                : entries 32, size 2.56 KiB, capacity 50 MiB, 1344
                           hits, 1395 requests, 0.963 recent hit rate, 14400
                           save period in seconds
Row Cache                : entries 0, size 0 bytes, capacity 0 bytes, 0 hits,
                           0 requests, NaN recent hit rate, 0 save period in
                           seconds
Counter Cache            : entries 0, size 0 bytes, capacity 25 MiB, 0 hits,
                           0 requests, NaN recent hit rate, 7200 save period
                           in seconds
```

```
Chunk Cache              : entries 30, size 1.88 MiB, capacity 221 MiB, 128
                           misses, 2738 requests, 0.953 recent hit rate,
                           866.987 microseconds miss latency
Token                    : (invoke with -T/--tokens to see all 256 tokens)
```

If this command nodetool hangs for over a minute or so, it means there's something wrong with your server's network configuration.

Stopping Cassandra

There's no Cassandra command to stop a running instance. If you started Cassandra in the foreground, press Control-C to stop Cassandra. If you started Cassandra in the background, you can stop the instance by using the Linux kill command. First, find the PID (Linux process ID) with the pgrep -f CassandraDaemon command and then kill it with the Linux kill command:

```
$ sudo pgrep -f CassandraDaemon
2284
```

Alternatively, you can simply run the ps command to get Cassandra's PID:

```
$ sudo ps auwx | grep cassandra
2284
```

Once you get the PID for Cassandra using one of the two methods I showed, you can kill the instance thus:

```
$ sudo kill 2284
```

Alternatively, you can kill the instance in a single step by running the following command:

```
$ sudo pkill -f CassandraDaemon
```

You'll notice that there's a script named stop-server in the $CASSANDRA_HOME/bin directory. However, the script doesn't stop anything! If you run it, it will suggest that you read the script before using it.

```
echo "please read the stop-server script before use"
# if you are using the cassandra start script with -p, this
# is the best way to stop:
```

```
# kill 'cat <pidfile>'

# otherwise, you can run something like this, but
# this is a shotgun approach and will kill other processes
# with cassandra in their name or arguments too:
# user='whoami'
# pgrep -u $user -f cassandra | xargs kill -9
```

You can create a simple shell script such as the following to shutdown Cassandra:

```
#!/bin/bash
CASS_PI
FID='ps -ef |grep CasandraDaemon |grep -v grep |awk '{ print $2 }"
if [[ "$CASSPID" == '']]
then
  echo Cassandra is NOT running
  else
    kill $CASS_PID
  fi
```

Starting and Stopping with the service Command

You can also start Cassandra as a service (Java server process) for a packaged installation. You'll find the startup scripts in the /etc/init.d directory. The service runs under the Cassandra user.

On a Debian system, the Cassandra service will automatically start following the installation of the software. You can check the service's status, stop, and restart it with following commands:

```
$ sudo service cassandra status
$ sudo service cassandra stop
$ sudo service cassandra start
```

When you run the start service command in a cluster with multiple nodes, on initial startup you must start each node one at a time, beginning with the seed nodes.

If you try to run the `service cassandra` command in a tarball installation, you'll receive an error:

```
# service cassandra status
● cassandra.service
   Loaded: not-found (Reason: No such file or directory)
   Active: inactive (dead)
#
```

Clearing Cassandra Data

Sometimes you may need to clear Cassandra data. You may need to remove just the data in the data directory, or from all default directories. I explain the procedures for removing the directories in this section.

The procedure to clear Cassandra data is similar in both package installations and standalone installations, differing only in the location of the default directories.

To clear the data from all default directories in a package installation, do the following:

```
$ cd install_location
$ rm -rf data/*
```

This will remove data from the default directories, including the commit log and the saved-caches directories.

To remove just the data directory, do the following:

```
$ sudo rm -rf data/data/*
```

You can remove all data in a package installation (from the default directories) by doing this:

```
$ sudo stop service cassandra
$ sudo rm -rf /var/lib/cassandra/*
```

Verifying the Cassandra Version

The output of the `cassandra` command tells you the Cassandra version. You can also find the version by running the `cassandra` command with the -v option:

```
$ ./cassandra  -v
3.9
$
```

The -v option simply prints the Cassandra version and quits. You can also execute the `show version` command to get version info:

```
cqlsh> show version;
[cqlsh 5.0.1 | Cassandra 3.7 | CQL spec 3.4.2 | Native protocol v4]
cqlsh>
```

The nodetool version command also shows the Cassandra version:

```
$ nodetool version
ReleaseVersion: 3.11.0
$
```

Configuring cqlsh

The cqlsh utility is highly configurable, and you can configure it via a dedicated configuration file or by choosing from several options at the command line. Let's review both ways of configuring cqlsh.

Configuring Through the cqlshrc Config File

You can configure various properties in the cqlshrc file. Cassandra provides a `cqlshrc.sample` file for you, and you can rename it to cqlshrc. The `cqlshrc.sample` file is located in the $CASSANDRA_HOME/conf directory. Here's a partial listing of the `cqlshrc.sample` file:

```
cql]
;; A version of CQL to use (this should almost never be set)
; version = 3.2.1

[connection]
;; The host to connect to
hostname = 127.0.0.1
```

```
;; The port to connect to (9042 is the native protocol default)
port = 9042

;; Always connect using SSL - false by default
; ssl = true

;; A timeout in seconds for opening new connections
; timeout = 10

;; A timeout in seconds for executing queries
; request_timeout = 10
...
```

Configuring by Specifying Options at the Command Line

You can specify several options for cqlsh at the command line. To take a simple example, you can specify a non-default location for the cqlsh config file, cqlshrc, with the –cqlshrc option:

```
$ cqlsh –cqlshrc $CASSANDRA_HOME/newconf
```

Finding the Versions

You can find the versions of not only Cassandra, but also cqlsh, CQL, and the native protocol, with the same VERSION command:

```
cqlsh> show version
[cqlsh 5.0.1 | Cassandra 3.9 | CQL spec 3.4.2 | Native protocol v4]
cqlsh>
```

Cqlsh Options

Cqlsh comes with numerous options to facilitate your work, and I describe the most useful options in the following sections.

Clearing the Screen

You can clear the screen by typing either clear or CLS at the cqlsh command line.

Running Commands from a File

Often, you'll want to run a set of commands one after the other. At times like this, you can simply store all the commands in a text file using a text editor such as vi or nano. For example, the following two lines are in a file named myfile:

```
use mykeyspace1
select * from employees
```

Once you create the text file with the cqlsh commands, you can invoke the commands with the source command, as shown here:

```
cqlsh> source  '/cassandra/test/myfile';
```

Putting Cassandra Through Its Paces

Now that you've installed, configured, started, and stopped the Cassandra instance, you've gotten over a major hurdle. To see how Cassandra functions, it's a good idea to create some test data at this point.

In this section, I'll connect to CQL shell and create a keyspace and a table and then I'll query from that table. Just follow along for now, and I explain the syntax and other interesting stuff in Chapters 4 and 5.

Connecting to the CQL Shell

You can connect to the CQL shell by typing cqlsh at the command prompt, as shown here:

```
$ ./cqlsh
Connected to Test Cluster at 127.0.0.1:9042.
[cqlsh 5.0.1 | Cassandra 3.9 | CQL spec 3.4.2 | Native protocol v4]
Use HELP for help.
cqlsh>
```

Note that the cluster to which cqlsh connects is the Cassandra cluster named Test Cluster, which happens to be the default name for a Cassandra cluster.

In this case, I didn't specify a Cassandra node that cqlsh ought to connect to, so it connects to the Cassandra instance running on localhost. If you're running the cqlsh command in a multi-node cluster, you can connect to a specific node in a cluster. To do this, you specify the hostname and port on the command line:

```
$ cqlsh 192.168.177.140 9160
```

Creating a Keyspace

As you'll learn shortly, to store data in a Cassandra table, you must first create a keyspace. Once you create this keyspace, you can create tables inside it for storing data.

Here's how you create a keyspace named testdata:

```
cqlsh> create keyspace testdata with replication = {'class' :
'SimpleStrategy', 'replication_factor' : 2};
cqlsh>
```

This example creates a keyspace named testdata with a replication level of 2. Never mind the syntax of the command; you'll learn all that in Chapter 4.

Creating a Table

Now that you have your keyspace ready, it's time to create your first Cassandra table! Before you issue the create table command, run the use testdata command so Cassandra can create the table in the testdata keyspace.

```
cqlsh> use testdata;
cqlsh:testdata>
cqlsh:testdata> create table quarters ( id int PRIMARY KEY, name text );
```

You can verify that Cassandra created your new table by running the describe tables command:

```
cqlsh:testdata> describe tables;
quarters
cqlsh:testdata>
```

Inserting Test Data

Let's insert some test data into the quarters table.

```
cqlsh:testdata> insert into quarters (id,name) VALUES(1, 'Spring');
cqlsh:testdata> insert into quarters (id,name) VALUES(2, 'Summer');
cqlsh:testdata> insert into quarters (id,name) VALUES(3, 'Fall');
cqlsh:testdata> insert into quarters (id,name) VALUES(4, 'Winter');
cqlsh:testdata>
```

Querying the Table

Let's query the quarters table.

```
cqlsh:testdata> select * from quarters;
 id | name
----+--------

  1 | Spring
  2 | Summer
  4 | Winter
  3 |    Fall
(4 rows)
cqlsh:testdata>
```

Getting the History of Your Commands

You can get a history of all your CQLSH as well as Nodetool commands by going to the directory named .cassandra under the /home directory of the user as which you run these commands.

```
ubuntu2:/home/samalapati/.cassandra$  ls -altr
total 16
-rw-------  1 samalapati samalapati    5 Mar  5 17:22 cqlsh_history
-rw-rw-r--  1 samalapati samalapati 3156 Mar 11 11:44 nodetool.history
...
```

Inside the hidden directory `.cassandra` you'll find two files: `cqlsh_history` and `nodetool.history`, which store the history of all the `cqlsh` and `nodetool` commands that you have run, respectively.

You now have a spanking new Cassandra single-node cluster. It's time to move on to a multi-node cluster in the next chapter.

Summary

You can install Cassandra in several ways: through the source, a binary package, or from a through a binary tar ball.

When dealing with Casandra for the first time, you can start with just a handful of configuration parameters, and learn to configure the rest of the properties as you learn more about Cassandra.

Learning how to run the CQL shell commands will help enhance your productivity. Go through the various Cassandra directories, such as `bin`, `conf`, `tools`, and `logs`, so you can familiarize yourself with the entire toolkit that Cassandra offers. This tour of the directories also helps you learn where Cassandra stores various things such as data, logs, snapshots, and various other artifacts.

It's a good idea to create start and stop scripts for your clusters.

Deploying a Cassandra Cluster

In Chapter 2, you learned how to install and configure a single-node Cassandra cluster. The idea behind starting with a simple one-node cluster was to get novices comfortable with the Cassandra terminology and learn the basics of starting and stopping Cassandra nodes.

The real strength of Cassandra, however, lies in its distributed architecture, so in this chapter, I show how to create and configure a multi-node cluster.

You'll learn how to create, start, and stop a multi-node cluster, both with a single data center and multiple data centers. The chapter also shows you how to create a multi-node Cassandra cluster in the cloud, in an Amazon Web Services environment.

Planning a Cluster Deployment

When you are planning the deployment of a Cassandra cluster, you must figure out both the number of nodes you want to start out with and the configuration of those nodes. For small development clusters, the configuration of the nodes isn't critical. However, choosing the right configuration for memory, CPUs, disk, and network is critical for a production deployment.

Using cassandra-stress for Planning a Production Deployment

Cassandra offers a great tool for stress testing a cluster before you start production operations on that cluster. The tool is named cassandra-stress, and I explain it in detail in Chapter 11, which deals with tuning Cassandra performance.

© Sam R. Alapati 2018
S. R. Alapati, *Expert Apache Cassandra Administration*, https://doi.org/10.1007/978-1-4842-3126-5_3

Choosing Memory

Regardless of whether you use virtual or dedicated hardware, you need to ensure that you have enough memory for your production Cassandra environments. Although Cassandra needs only a minimum of 8GB of RAM, the server should have at least 64GB to 512GB of RAM.

There is no "ideal" amount of RAM for a Cassandra node. The amount of memory depends on the amount of data that the node will process. Remember that data writes first go to tables in memory (memtables) and from there to the SSTables that live on disk.

If a Cassandra node has too little memory, it'll end up with smaller memtables, meaning that the database must flush a larger number of SSTables to disk. This means that queries will need to perform the more expensive disk I/O to read a large number of files on disk. The bottom line is that the more RAM you can purchase, the better.

Choosing CPUs

Cassandra is optimized for writing, so the CPU is the limiting factor for performance. Workloads that insert data are CPU-bound before they become memory-bound.

DataStax recommends that you use dedicated hardware, with an ideal CPU count of 16 processors.

Network Considerations

Cassandra is a distributed database, and therefore the network needs to transmit vast chunks of data for both the read and write activity and for data replication. The recommended bandwidth for the network is 1GB or higher.

Choosing Storage

Optimal storage choices include the type of storage, such as SAN and other storage types, as well as the size of the storage capacity. Understanding how Cassandra uses storage also helps in making optimal storage choices.

Cassandra can make smart use of the high IOPs offered by SSD. Let's say you have a set of tables whose data you only read rarely. You can use SSDs for frequently used column families to increase the I/O speed and use normal storage drives for the infrequently accessed data.

NFS, SAN, and NAS Not Advisable

DataStax recommends that you not use SAN, NDFS, or NAS storage for Cassandra environments. DataStax also recommends that you not use SAN or NAS storage for an on-premise Cassandra cluster.

SAN isn't advisable due to the following reasons:

- SAN's return on investment isn't attractive in a Cassandra environment, as you keep scaling the cluster.

- Since Cassandra's I/O is often higher than the ability of the array controller, SAN becomes a bottleneck, or even worse, becomes a single point of failure.

- Despite SSD and high-speed networks, SAN adds latency to Cassandra's operations.

- SAN transport, occurring simultaneously with Cassandra traffic, can saturate your network, causing problems to all network users.

If you *must* use SAN, you need expertise in investigating issues such as SAN fiber saturation.

Network-attached storage (NAS) devices are not recommended because they cause network bottlenecks due to high I/O wait times for reads and writes. The bottlenecks may stem from router latency as well as issues with the network interface cards (NICs).

Finally, DataStax does not recommend Network File System (NFS) storage due to its inconsistent behavior during the deletion and moving of files.

How Cassandra Uses Disk Storage

The key to understanding Cassandra's storage requirements is to understand how and when Cassandra writes data to disk. Cassandra writes data to disk under the following scenarios:

- When it writes to the commit log

- When it flushes memtables to SSTable data files

- When it periodically compacts the SSTables (compaction temporarily increases disk usage)

71

Compaction and Storage Requirements

Compaction requires enough free space on disk to complete the compaction work. The storage requirements for compaction depend on the size of all your SSTables, as well as the compaction strategy you adopt.

In the worst case, when you use the compaction strategy called SizeTieredCompactionStrategy, you need free storage space that is 50 percent of the sum of all the SSTables that the database is compacting.

I'll explain the storage requirements for compaction in Chapter 11.

Estimating Usable Disk Capacity

Cassandra uses disk storage mostly for storing the commit log and the data directories, the latter of which stores the SSTable data. In addition to this, it requires some free storage to perform compaction work. Ideally, you should store the commit log on a different storage drive from where you store the data directories.

You can use the following method to calculate the total usable storage capacity of your cluster.

1. Calculate the total raw capacity of the physical disks.

    ```
    raw-capacity - size of disk * number of disks per server
    Example: 12 disks * 7.2 TB = 86.4 TB
    ```

2. Calculate the usable disk space by deducting 10% for formatting overhead.

    ```
    formatted disk space - raw capacity * 0.9
    Example: 86.4 * 0.9 = 76.76 TB
    ```

3. Since Cassandra requires disk storage for compaction and repair operations, you can't use all of a disk's formatted disk space for the commit log and the data directories. DataStax recommends that you allocate only 50-80 percent of a storage drive's capacity for storing data and leave the rest for compaction activity.

    In this case, I have about 77TB of formatted disk space, so I can use anywhere between 38.5TB to 62TB of it for storing data (SSTables + commit log).

Choosing Production Settings for a Linux Server

As mentioned earlier, this book uses Linux servers for running a Cassandra cluster. You can run Cassandra on most Linux versions. I use an Ubuntu 16.04 LTS server for all the work in this book.

Java Version

You must use the latest version of Java, and you can use either the Oracle Java Platform or OpenJDK. I use the Oracle Java Platform, Standard Edition 8 JDK.

Linux Server and Kernel Settings

To optimize an Apache Cassandra installation, follow DataStax's recommendations, which I explain in the following sections.

Synchronize the Clock and Enable NTP

Cassandra overwrites a column only if there's a newer version with a more recent timestamp. Synchronizing the clocks on all the nodes of a cluster is thus critical. You can use NTP (Network Time Protocol) or another method for synchronizing the clocks.

You must set up one of the cluster's servers as an NTP server if your cluster doesn't have access to the Internet. Synchronize the network time on all the cluster's nodes by enabling the NTP daemon through editing the /etc/sysconfig/ntpd file.

Disable the zone_reclaim_mode on NUMA Systems

To avoid potential performance problems, you must disable the zone_reclaim_mode. You can check if the mode is enabled by doing this:

```
$ cat /proc/sys/vm/zone_reclaim_mode
0
$
```

If the output is not a zero, disable the mode with the following command:

```
$ echo 0 > /proc/sys/vm/zone_reclaim_mode
```

TCP Settings

You need to bump up the default TCP settings, which are adequate only for a small setup. For a busy production environment, with hundreds or thousands of concurrent connections, you need to set the following values for the TCP parameters by editing the /etc/sysctl.conf file:

```
net.core.rmem_max = 16777216
net.core.wmem_max = 16777216
net.core.rmem_default = 16777216
net.core.wmem_default = 16777216
net.core.optmem_max = 40960
net.ipv4.tcp_rmem = 4096 87380 16777216
net.ipv4.tcp_wmem = 4096 65536 16777216
```

You must reboot the Linux server for the new TCP settings to go into effect. If you want to change the settings without restarting the server, you can issue the following command:

```
$ sudo sysctl -p /etc/sysctl.conf
```

User Resource Limits

You must set the following limits on various user resources, in the /etc/security/limits.conf file:

```
<cassandra_user> - memlock unlimited
<cassandra_user> - nofile 100000
<cassandra_user> - nproc 32768
<cassandra_user> - as unlimited
```

In addition, you also need to include the following setting:

```
vm.max_map_count = 1048575
```

If you are using a Red Hat-based Linux server, you must also set the following nproc limit in the /etc/security/limits.d/90-nproc.conf file:

```
cassandra_user   - nproc 32768
```

PAM Security Settings

For many Linux versions, you must enable the `pam_limits.so` module. You do this by uncommenting the following line in the `/etc/pam.d/su` file:

```
session  required  pam_limits.so
```

Setting the Java Heap Size

Tuning Java garbage collection (GC) is the single biggest thing you can do to tune a Java application. Cassandra runs in a JVM, so setting the right heap size is critical for performance.

There are several Java garbage collection algorithms. The most common is the Concurrent-Mark Sweep (CMS) garbage collector. However, in newer releases of Java, the G1 (Garbage First) garbage collector is frequently the recommended garbage collection algorithm.

Here's what you need to know about choosing the Java garbage collector:

- CMS involves more time and effort to tune it, whereas G1 is easy to configure and mostly tunes itself.

- CMS is more appropriate for fixed workloads and G1 for workloads that are frequently changing, with the cluster running different processes.

- For smaller environments that use a heap size that's less than 14GB and where latency is a major concern, the recommended collector is CMS.

- For environments that use a heap size of between 14GB-64GB, G1 is better than CMS. Unlike CMS, which stops all application threads during garbage collection, the G1 garbage collector's threads scan the heap's regions and the collector performs the heap compaction while the application threads are running.

- You must set the heap size to something between 0.25 and 0.5 of the total RAM on the server running the Cassandra instances.

Chapter 11 explains Java garbage collection in detail and shows how to review GC logs to configure the optimum heap size for Cassandra.

Disabling Swap

To avoid a high latency due to swapping under low free-memory conditions, you should disable swap entirely. Here is how you do it:

```
$ sudo swapoff -all
```

For this change to survive a reboot, you must also remove the swap file related entries from the /etc/fstab file.

By default, most Linux operating systems come with a default swappiness setting of 60. If you set swappiness to zero, Linux will avoid using the disk unless it runs out of memory, whereas setting it to 100 means that the OS will instantly swap programs to disk. As you can tell, a setting of 60 means that the OS will use the swap file on disk often, starting from the time when the memory usage reaches around half the OS RAM allocation. If you turn swappiness down to 10, for example, the OS will use the swap file on disk only when the RAM usage is around 90 percent.

The Linux administrator can change the system swappiness value by adding the following to the /etc/sysctl.conf file:

```
vm.swappiness=10
```

The administrator must reboot the server for the new swappiness setting to take effect. There is no fixed rule on how low you must set the swappiness level; many experts recommend setting it to 1.

Setting the Limits

Limit the cluster resources that users can utilize by setting shell limits. You can do this by editing the /etc/security/limits.conf file, which dictates the limits on how users can use resources. You use the limits.conf file to configure "soft" and "hard" limits on important operating system properties such as file sizes, the stack size, and the priority levels (niceness) of processes.

Add the following lines to your /etc/security/limits.conf file:

```
soft nofile    32768
hard nofile    32768
hard nproc     32768
soft nproc     32768
```

The nofile attribute limits the number of open descriptors per user process and nproc specifies the maximum number of processes. The *soft* limit settings connote warnings and the *hard* limit settings are the actual resource limits.

The Java Hugepages Setting

By default, in most new Linux distributions, the transparent hugepages feature is enabled, meaning that when handling transparent hugepages, the kernel allocates memory in chunks sized 2MB each, rather than 4K. Sometimes, when dealing with applications that allocate memory in 4K-sized pages, the server performance takes a hit when the kernel needs to defragment the large 2MB pages, which are fragmented by many tiny 4K pages.

You can avoid the performance hit due to the defragmentation by disabling defrag for hugepages, as shown here:

```
$ echo never | sudo tee /sys/kernel/mm/transparent_hugepage/defrag
```

Installing PDSH

Since you'll be managing multi-node clusters, it's a good idea to acquire a tool that can help you simultaneously run commands or send files to multiple nodes. You can make cluster administration easy by using a tool such as pdsh to simultaneously run commands on your entire cluster. I show you how to download, install, and use this tool.

The pdsh utility is a variant of the rsh command and is a high-performance parallel shell utility. Whereas rsh lets you run commands on a single remote host, pdsh lets you simultaneously run commands on multiple remote servers.

When you need to issue the same command across all the node of a Cassandra nodes, simply issue the command from a single server using pdsh, and that'll executes the command across the cluster.

You can issue several types of Linux commands across the cluster using pdsh, including commands that view the contents of a file.

You can use pdsh by issuing commands at the command line, or by running the tool interactively. When run interactively, pdsh prompts you for commands and executes them when a carriage return occurs. You can also specify your commands in a file.

The pdsh distribution also includes a parallel remote copy utility named pdcp, which copies files from a local host to a group of remote hosts in parallel.

You can install pdsh in the following manner:

```
# rpm -Uvh http://download.fedoraproject.org/pub/epel/6/i386/
epel-release-6-8.noarch.rpm
# yum install pdsh
```

Using pdsh to perform remote operations is straightforward. Here's an example that shows how to check the date on all nodes in a cluster, by running a single command from any node in the cluster:

```
# pdsh -w "all_nodes" date
```

The parameter `all_nodes` points to a file that lists all the nodes in the cluster. You can also exclude some servers if you wish by specifying the appropriate option when issuing a `pdsh` command.

Initializing a Cassandra Multi-Node Cluster (Single and Multiple Datacenters)

In Chapter 2, you learned how to configure a single-node Cassandra cluster. In this chapter, you'll learn how to configure a multi-node Casandra cluster, first with a single datacenter and later with multiple datacenters.

In a multiple-node cluster, Cassandra automatically discovers nodes, so to set up a Cassandra cluster, you simply install Cassandra on all the nodes, as I show in the following section. Once you install Cassandra, you start the Cassandra instances, and they automatically form a cluster. All you need to do is to let each node know the IP addresses of the rest of the nodes; that's it!

As you can recall from Chapter 1, a datacenter is nothing but a grouping of nodes and represents a set of nodes that have the same replication properties. A datacenter can be logical or physical.

In the discussion that follows, I create a Cassandra cluster from scratch. If you are instead turning a single-node cluster into a multi-node cluster, you must first stop the Cassandra server and clear the data, as explained in Chapter 2.

The steps I show here will enable to you to install, configure, and run a multi-node cluster using a single datacenter. In order to set up a multi-node cluster with multiple datacenters, you follow the same steps but configure multiple datacenters in the `cassandra-rackdc.properties` file, as I explain later in this section.

Prerequisites

Before you can get your multi-node cluster going, you must take care of some prerequisites, as explained in the following sections.

Configuring Firewall Port Access

When you have a single-node cluster, the firewall access ports aren't a big deal. In a multi-node cluster, you must ensure that if a firewall is running on the nodes hosting the Cassandra cluster, you must open several ports, including some Cassandra ports, to enable the ports to communicate among themselves.

If you forget to open the ports after starting Cassandra on one of these nodes, the node won't join the cluster and will act as a standalone Cassandra instance.

There are three sets of ports you must open: public ports, Cassandra inter-node ports, and Cassandra client ports.

Public Ports

Port Number 22, which serves as the SSH port

Cassandra Inter-Node Ports

Port 7000: For Cassandra inter-node cluster communication
Port 7001: For Cassandra SSL inter-node cluster communication
Port 7199: For Cassandra JMX monitoring

Cassandra Client Ports

Port 9042: Cassandra native client port
Port 9160: Cassandra client port (Thrift)

Selecting a Name for the Datacenter

Before you create the multi-node cluster, select a naming convention for each datacenter and rack in the cluster. The datacenter name is a required parameter and serves to ensure that nodes that don't belong to this datacenter don't attempt to join it.

Tip   Once you assign a name for a datacenter, you can't change it later.

In this case, I select *datacenter1* as the name of the datacenter.

Gathering the IP Addresses for All of the Nodes

You need to get the IP addresses for all the nodes in the cluster. In this case, I have six nodes and their IP addresses are

```
node0  192.168.177.132
node1  192.168.177.133
node2  192.168.177.134
node3  192.168.177.135
node4  192.168.177.136
node5  192.168.177.137
```

The six-node cluster will span two racks and there's one datacenter.

Selecting the Nodes to Serve as Seed Nodes

A seed provider is one of the nodes in the cluster that helps Cassandra nodes to find each other and learn the topology of the ring. This is a required parameter for a multi-node cluster.

You specify the seed nodes(s) for a cluster by configuring the seed_provider parameter in the cassandra.yaml file.

You configure the seed_provider parameter by setting the value for the seeds attribute of this parameter, as shown here:

```
seed_provider:
    - class_name: org.apache.cassandra.locator.SimpleSeedProvider
      parameters:
      - seeds: "192.168.177.132,192.168.177.135"
```

You provide the list of seed nodes as a comma-delimited set of IP addresses ("<ip1>, <ip2>, <ip3>").

You can make do with a single seed node per datacenter, but the best practice is to have more than one seed node. In this case, I chose to have two seeds, so my seeds attribute has the value

```
"192.168.177.132,192.168.177.135"
```

Configuring the Cluster

You configure the cluster properties in the all-important `cassandra.yaml` file, which you can find in the `$CASSANDRA_HOME/conf` directory.

As I mentioned in Chapter 2, there are hundreds of parameters you can configure in this file, but listing a bunch of them here doesn't do you a whole lot of good. I therefore list a minimal set of properties to start the new cluster, meaning that the rest of the parameters will be using their default values. As you move through the rest of the chapters, you'll find explanations for all the configuration parameters you can specify in the `cassandra.yaml` file.

Set the following properties in the `cassandra.yaml` file for each node of the cluster. Alternatively, you can set the properties in one node and copy the file over to the nodes using the pdsh tool described earlier.

The num_tokens Property

The `num_tokens` property defines the number of tokens Cassandra assigns to a specific node. The higher the number of tokens relative to the rest of the nodes, the greater the amount of data this node will store. Since ideally all nodes are of equal size, you want all nodes to have the same number of tokens.

Note   The `initial_token` property is a legacy parameter that you must leave alone. If you specify the `initial_token` property, it'll override the `num_tokens` property.

The `num_tokens` property helps create *virtual nodes* or *vnodes*, which help break up the token range into many small ranges. Each Cassandra node is then assigned a set of the vnodes. Cassandra calculates the token ranges for each cluster node based on that node's `num_tokens` value.

Vnodes are especially useful when dealing with clusters that have machines with different configuration and capacities. You can assign more vnodes to machines that have more computing resources available by specifying a higher value for the `num_tokens` property for these nodes when compared to the rest of the nodes.

Since vnodes help break up the token ring into multiple smaller ranges, they help load cluster operations more evenly across the nodes, and thus speed up several operations such as bootstrapping new nodes.

The default as well as the recommended value for the num_tokens property is 256.

The –seeds Property

I've already explained the -seeds property, which requires you to specify the internal IP address for the node(s) you've selected as the seed node(s0).

The listen_address Parameter

The listen_address parameter refers to the IP address of the node. This is the network address to bind to and tell other Cassandra nodes to connect to. You can set this or leave it alone. If you don't set this property, Cassandra gets the local address from the host, but sometimes it's unable to get the correct address, in which case you must provide the listen_address in the cassandra.yaml file.

You must not set the value 0.0.0.0 for the listen_address property.

The rpc_address and broadcast_rpc_address Properties

The rpc_address property specifies the address to bind the Thrift RPC service and native transport server. You can leave the rpc_address property blank, in which case Cassandra picks it up based on the hostname you've configured for the node.

Unlike in the case of the listen_address property, you can specify the value 0.0.0.0 for the rpc_address property, in which case you must also set the value for the property broadcast_rpc_address:

```
broadcast_rpc_address: 192.168.177.135
```

The endpoint_snitch Option

In a Cassandra cluster, a *snitch* serves two functions:

1. It tells Cassandra about the network topology so it can efficiently route its requests.

2. It enables Cassandra to spread the data copies (replicas) around the cluster, thus avoiding correlated failures. Cassandra uses datacenters and racks to logically group a cluster's nodes. It tries its best not to store multiple replicas of the same piece of data on a single rack.

Cassandra offers a half dozen snitches, but for production environments, the go-to option is the GossipingPropertyFileSnitch.

When you select the GossipingPropertyFileSnitch option, you specify the datacenter and the rack in the cassandra-rackdc.properties file on that node. Cassandra then propagates this information to the other nodes via gossip.

The auto_bootstrap Property

The default value for the auto_bootstrap property is true, and the parameter is not present in the cassandra.yaml file. This parameter makes new non-seed nodes migrate the data to themselves. When you're initializing a new cluster with no data, add the following property:

```
auto_bootstrap=false
```

With the minimal set of configuration properties I've listed here, my cassandra.yaml file looks as follows:

```
cluster_name: "MyCluster'
num_tokens: 256
    seed_provider:
        - class_name: org.apache.cassandra.locator.SimpleSeedProvider
          parameters:
          - seeds: "192.168.177.132,192.168.177.135"
listen_address:
endpoint_snitch: GossipingPropertyFileSnitch
```

Configuring the Datacenter and Rack Names

In my cluster, I have a single datacenter and a single rack. I must specify the datacenter and rack names in the cassandra-rackdc.properties file.

```
# indicate the rack and dc for this node
dc=DC1
rack=RACK1
```

I need to do this on all six nodes of the cluster.

At this point, the installation and configuration of Cassandra is complete. I'll show you how to start up this cluster in the section "Starting and Stopping the Multi-Node Cluster."

Initializing the Cluster with Multiple Datacenters

The previous discussion shows how to set up a Cassandra cluster that uses a single datacenter and a single rack. Configuring a cluster that uses multiple datacenters and racks is just as easy! You follow the same steps, except that you configure multiple datacenters and racks in the `cassandra-rackdc.properties` file.

In the single datacenter case, I specified the following properties in the `cassandra-rackdc.properties` file:

```
dc=DC1
rack=RACK1
```

I want to set up two datacenters, each with three of my six Cassandra nodes. In order to do this, I configure the `cassandra-rackdc.properties` file in the following way.

In the `cassandra-rackdc.properties` file for the first three nodes, 192.168.177.32, 192, 168,177.33, and 192.168.177.34, I specify the following values for the datacenter and rack:

```
dc=dc1
rack=rack1
```

For the other three nodes of my six-node cluster, I specify the following values:

```
dc=dc2
rack=rack2
```

That's all you need to do to create a cluster with multiple datacenters (I changed the rack to rack2, but I really didn't need to do this).

Starting and Stopping the Multi-Node Cluster

You have your cluster all configured and ready to go, except that you need to start it up. To start the cluster, first start the two seed nodes, one after the other. Once you do this, start the other four nodes one after the other. Since this is a tarball installation, I start the nodes with the following command:

```
$ $CASSANDRA_HOME/bin/cassandra
```

You can now check that the ring is running by doing this:

```
# CASSANDRA_HOME/bin/nodetool status
```

To be ready for usage, all six nodes should be showing the status UN (Up Normal). You can use the following set of scripts to start and stop your multi-node cluster.

Script for Starting the Cluster

You can't manage a cluster without scripts. You can write scripts that are more sophisticated than what I offer here, but this one does the job in helping start a cluster.

In this example I have three nodes in the Cassandra cluster.

```
#!/bin/bash
SERVERS="
192.168.177.131
192.168.177.132
192.168.177.133"
for SERVERNAME in $SERVERS; do

    sleep 30

    echo Starting node $SERVERNAME...
    sudo -u cassandra ssh $SERVERNAME "usr/share/cassandra/bin/cassandra"
done
```

The sleep command is there to provide a short gap before starting each server. In a production cluster, running this script without the sleep command may cause issues.

Scripts for Stopping the Cluster

You know that you can use the cassandra command to start a cluster but not to stop it. You must kill the PID of the running Cassandra instance to stop the instance. To automate the stopping of the Cassandra instance for a set of nodes, you can use the following strategy:

- Use the first script to iterate through the list of servers and run your stop script.

- Use a second script that stops the *cassandra* service by killing the PID of the Cassandra instance.

Here are the two scripts.

Script 1

Use this script to call the `cassandra-kill.sh` script.

```
#!/bin/bash
SERVERS="
192.168.177.131
192.168.177.132
192.168.177.133"
for SERVERNAME in $SERVERS; do
    echo Starting node $SERVERNAME...
    sudo -u cassandra ssj $SERVERNAME "/usr/share/cassandra/bin/cassandra"
done
```

Script 2 (`cassandra-kill.sh`)

You can create a simple shell script such as the following to shut down a Cassandra node:

```
#!/bin/bash
CASS_PID='ps -ef |grep CasandraDaemon |grep -v grep |awk '{ print $2 }"
if [[ "$CASSPID" == '']]
then
  echo Cassandra is NOT running
  else
    kill $CASS_PID
  fi
```

The Startup Process of the Nodes in a Cluster

In Chapter 2, I showed the startup process for a single node. In this chapter, I show how to create a cluster with multiple nodes. Note how when you start the first node, it shows that it's ready for work and shows the other nodes joining the cluster. In this case, I am reviewing the startup messages for the node 192.168.177.132.

```
INFO  14:12:49 Node /192.168.177.132 state jump to NORMAL
INFO  14:12:49 Waiting for gossip to settle before accepting client
      requests...
INFO  14:12:57 No gossip backlog; proceeding
```

```
INFO  14:12:58 Starting listening for CQL clients on /0.0.0.0:9042
      (unencrypted)...
INFO  14:12:58 Binding thrift service to /0.0.0.0:9160
INFO  14:12:58 Listening for thrift clients...
INFO  14:12:59 Handshaking version with /192.168.177.135
INFO  14:12:59 Scheduling approximate time-check task with a precision of
      10 milliseconds
INFO  14:12:59 Handshaking version with /192.168.177.135
INFO  14:13:01 Node /192.168.177.135 has restarted, now UP
INFO  14:13:01 Updating topology for /192.168.177.135
INFO  14:13:01 Updating topology for /192.168.177.135
INFO  14:13:01 InetAddress /192.168.177.135 is now UP
INFO  14:13:01 Handshaking version with /192.168.177.135
INFO  14:13:01 Node /192.168.177.135 state jump to NORMAL
```

Similarly, when you bring down the second node (or it crashes), the messages from the first node show that information:

```
INFO  14:13:42 Handshaking version with /192.168.177.135
INFO  14:19:14 InetAddress /192.168.177.135 is now DOWN
```

Common Errors When Starting Out

You could run into numerous errors when running a distributed database such a Cassandra. I want to note a pair of common issues and how to overcome them.

Change in a Node's IP Address

When you're running Cassandra on a virtual machine, sometimes the node's IP address can change. When this happens, the rest of the nodes of course can't connect to this node and it'll show up as a down node.

The fix for the change in an IP address is simple. You just need to edit the cassandra.yaml file for this node and change the old IP address to the new IP address everywhere you've set the IP address of the node, such as the listen_address configuration property.

A Schema Version Mismatch

Sometimes you'll run into an error when creating a keyspace or a table, such as the following, where Cassandra complains about a version mismatch:

```
cqlsh> create keyspace mykeyspace2
    ... with replication = {'class': 'NetworkTopologyStrategy',
    'datacenter1' :2}
    ... and durable_writes = false;
Warning: schema version mismatch detected, which might be caused by DOWN
nodes; if this is not the case, check the schema versions of your nodes in
system.local and system.peers.
OperationTimedOut: errors={'192.168.177.135': 'Request timed out while
waiting for schema agreement. See Session.execute[_async](timeout) and
Cluster.max_schema_agreement_wait.'}, last_host=192.168.177.135
cqlsh>
```

Cassandra will create the keyspace or table despite this message. When a schema disagreement occurs, follow these steps.

1. Run the nodetool describecluster command.

```
$ sudo nodetool describecluster
Cluster Information:
        Name: Test Cluster
        Snitch: org.apache.cassandra.locator.DynamicEndpointSnitch
        Partitioner: org.apache.cassandra.dht.Murmur3Partitioner
        Schema versions:
                UNREACHABLE:27a8739d-28ac-34b7-b738-1b84859866cf:
                [192.168.177.135]
                282bdefc-9643-3fa5-b03a-4b9894cabb29: [192.168.177.132]
$
```

2. Restart the unreachable node(s).

3. Run the nodetoool describecluster command again, and ensure that all nodes have the same version number. The output of the command must show a single schema version for all nodes in the cluster.

Note   If you have several mismatched schemas (three or more), you need to stop the nodes of a given schema and let the other settle, and then restart the nodes one by one. This type of a situation occurs occasionally in multiple DC clusters.

Keyspaces with Different Settings

If the keyspaces of the nodes have different settings, you'll notice the following:

```
$ sudo nodetool status
```

   Datacenter: datacenter1

```
========================
Status=Up/Down
|/ State=Normal/Leaving/Joining/Moving
--  Address         Load     Tokens       Owns     Host ID         Rack
DN  192.168.177.128  114.06 MiB  256          ?        99c43633-c691-4dee-
b7af-35bc6e74dd67   rack1
UN  192.168.177.132  123.84 MiB  256          ?        b0ade950-937a-457c-
95eb-d3032897eeb1   rack1
Note: Non-system keyspaces don't have the same replication settings,
effective ownership information is meaningless
$
```

Node Is Down

If one of the nodes is down, you will see the following:

```
$ sudo nodetool describecluster
Cluster Information:
        Name: Test Cluster
        Snitch: org.apache.cassandra.locator.DynamicEndpointSnitch
        Partitioner: org.apache.cassandra.dht.Murmur3Partitioner
        Schema versions:
                44ed2562-e330-3030-af87-89cde4aa8992: [192.168.177.135]
                UNREACHABLE: [192.168.177.132]
#
```

Run the nodetool status command to check the status of the two nodes in your cluster:

```
$ sudo nodetool status
Datacenter: datacenter1
=======================
Status=Up/Down
|/ State=Normal/Leaving/Joining/Moving
--  Address          Load        Tokens       Owns (effective)  Host ID          Rack
UN  192.168.177.132  276 KiB     256              52.2%                   b0ade950-
937a-457c-95eb-d3032897eeb1   rack1
UN  192.168.177.135  296.85 KiB  256              47.8%                   99c43633-
c691-4dee-b7af-35bc6e74dd67   rack1
#
```

Running Cassandra on Amazon EC2

Many organizations and individuals run Cassandra clusters in the public cloud. You can run Cassandra on Microsoft Azure, Google Cloud, and Amazon Web Services (AWS). In this section, I walk you through the creation of a Cassandra cluster on AWS.

When you install Cassandra on Amazon EC2, you create the instances using an API for a supported platform such as Ubuntu 16.04 LTS and ensure that you get the AMI (Amazon Machine Image) from a trusted source. Once you download the AMI and get the server running, the Cassandra installation process is similar to that in Chapter 2.

Using Trusted AMIs

An AMI is a virtual appliance that you use to create a virtual machine within the Amazon Elastic Compute Cloud ("EC2"). An AMI is a machine template using which you can create new servers in an AWS cloud.

You must use only AMIs from a trusted source, such as the following:

- Ubuntu Amazon EC2 AMI locator

- Debian AmazonEC2image

- CentOS-6 images on Amazon's EC2 Cloud

Using untrusted sources for your AMIs will create a security risk; they will also perform slower due to the way they configure the EC2 installation.

Setting Up the AWS Instances for Cassandra

Before you do anything on AWS, you must have an account. Therefore, if you don't have one, create one now.

Once you've got your AWS account squared away, follow the steps shown in the next few sections to create the Cassandra cluster that runs on AWS EC2 virtual machines, called EC2 instances.

Starting the EC2 Instance Creation

In the AWS dashboard, click the EC2 logo, under the Compute section.

Selecting the AWS Region

Select an appropriate region for launching your instances. For example, if you're in the US, you may want to select the North Virginia region.

Creating the EC2 Instances

Under the Create Instance section, you'll see a "Launch Instance" button. Click it to start the Launch Instance wizard.

This is the crucial step where you create the EC2 virtual machines. You can select an OS for the machines via different methods, such as downloading an Amazon Machine Image (AMI). You can even provide your own "gold image" if you have one. To keep things simple here, let's run a cluster on Ubuntu servers, so select the Quick Start menu.

The Quick Start menu has six steps that you need to go through to create your EC2 instances.

1. *Choose an Amazon Machine Image (AMI)*: Select the Ubuntu server (Ubuntu 14.04 LTS).

2. *Choose Instance Type*: Use local SSDs and not EBS storage. This is a test cluster, so choose the m3 size. The m3 size is not from the free-tier, so it costs a little bit. The free tier instances are too tiny for learning much about Cassandra. Your instance type choice is m3.large, a machine with 2 cores, 7.5GB RAM, and a 32GB SSD storage.

3. *Configure Instance Details*: This step is where you specify the number of EC2 instances. Configure a three-node cluster, so put down 3 as the value for the *Number of instances* property and leave the rest of the properties (shutdown behavior, etc.) at the default settings.

4. *Add Storage*: You don't need any additional storage at this point so just move on to the next step.

5. *Tag Instance*: Tagging instances with a key-value pair helps you in sorting and finding instances easily, but you can skip this step because you have just three EC2 nodes.

6. *Configure Security Group*: This step lets you configure a security group, which is a set of firewall rules to control traffic to the instances. In this step, do the following:

   • Select the *Create a new Security Group* option.

   • Name the security group as MySecurityGroup and add four inbound rules, as shown here:

Type	Protocol	Port Range	Source
SSH	TCP	22	0.0.0.0/0 (allow from anywhere)
Custom TCP Rule	TCP	7000-7001	0.0.0.0/0
Custom TCP Rule	TCP	7199	0.0.0.0/0
Custom TCP Rule	TCP	9042	0.0.0.0/0
Custom TCP Rule	TCP	9160	0.0.0.0/0

These ports are the same ports described in the "Configuring Ports" section earlier in this chapter.

7. In the last step of instance creation, Review Instance Launch, review your instance selections and press the Launch button so AWS can create the instances for you.

8. Once AWS starts the instances, it asks you to select the key pair that you'll be entering when logging into the new instances:

   Select an existing key pair or create a new key pair

The key pair is a combination of a public key that AWS stores and a private key file that you store. You use the key pair to securely log into your instances. On the new Ubuntu servers, your private key file enables you to securely SSH into your instances.

Tip   Store your private key file (`.pem` file) securely, since losing it means that you need to terminate all the instances and start again from the beginning.

Create a new key pair by providing the key pair name *mykeypair* and click "Download key pair" to download the private key file. At this point, all the EC2 instances are running, and billing for all usage starts right now. It is a good idea to shut down your instances when you're done working with Cassandra so you don't incur additional charges when you aren't utilizing your test cluster!

You can view the instances now by clicking the Instances tab in the EC2 Dashboard. All three instances will show the "running" status under the Instance State column. By selecting any of the three instances, you can get the instance description, including the public IP address for that instance. You can then launch a Putty session using the IP address for the instance.

Installing Cassandra

Now that you have your AWS EC2 instances running, it's time to install Cassandra. Follow these steps to install, configure, and start Cassandra on the AWS EC2 instances.

You learned in Chapter 2 that you can install Cassandra as a service or install it from a binary tar ball. Since you have already learned how to install from the tarball, it's a good idea to learn how to install Cassandra as a service. Here are the steps you must follow to install Cassandra as a service.

1.   You can get the Cassandra Debian package from Apache itself or from DataStax. Use the DataStax repository here, by doing the following:

```
$ echo "deb http://debian.datastax.com/community stable
main"| sudo tee -a /etc/apt/sources.list.d/cassandra.
sources.list
```

2. Run the apt-get update command.

   ```
   $ sudo apt-get update
   ```

3. If you receive any errors about not having the public key for the DataStax repository, you need to add the DataStax public repo key as shown here, and rerun the apt-get update command.

   ```
   $ curl -L http://debian.datastax.com/debian/repo_key |
   sudo apt-key add-
   $ sudo apt-get update
   ```

4. Install the Cassandra binaries.

   ```
   $ sudo apt-get install cassandra
   ```

   Unlike in the binary tarball installation method shown in Chapter 2 (for a single instance), installing Cassandra as a service automatically starts the Cassandra instance. If you now issue the command sudo service cassandra status, it'll show that cassandra is running on this node.

5. The next step is to repeat the previous three steps on the rest of the EC2 instances. Once you do this, you can run the nodetool status command to check the status of the Cassandra instances.

   ```
   $ sudo nodetool status
                           Datacenter: datacenter1
   =======================
   Status=Up/Down
   |/ State=Normal/Leaving/Joining/Moving
   --  Address    Load    Tokens    Owns    Host ID      Rack
   UN  192.168.177.132  123.84 MiB  256            ?
   b0ade950-937a-457c-95eb-d3032897eeb1   rack1

   Note: Non-system keyspaces don't have the same replication
   settings, effective ownership information is meaningless

   $
   ```

None of the three Cassandra nodes are communicating among themselves yet. Let's enable the inter-node communications next.

Configuring the Cassandra Cluster

On all three EC2 instances, edit the `cassandra.yaml` file and add the following properties:

```
cluster_name: 'My AWS Cluster'

seeds: "192.168.177.132"
broadcast_address: 192.168.177.132
listen_address:
```

These are the same properties that I explained earlier when creating the six-node Cassandra cluster. As in the case of that cluster, I decided not to specify a value for the `listen_address` property and therefore I must specify a value for the `broadcast_address` property.

Note   Users often see the EC2 snitch and use it, since it's designed for use in an AWS cluster. However, the recommended snitch to use is `GossipingPropertyFileSnitch`.

Once you edit the `cassandra.yaml` file, restart the Cassandra service on all three nodes, making sure to remove all the system data:

```
$ sudo service cassandra stop
$ sudo rm -rf /var/lib/Casandra/data/system/*
$ sudo service cassandra start
```

Running the `nodetool status` command shows that all three nodes are running now, and that the new three-node AWS EC2-based Cassandra cluster is ready for use.

```
INFO  22:20:27 Handshaking version with /192.168.177.135
INFO  22:20:27 Node /192.168.177.135 has restarted, now UP
INFO  22:20:27 InetAddress /192.168.177.135 is now UP
INFO  22:20:27 Node /192.168.177.135 state jump to NORMAL
INFO  22:20:27 Node /192.168.177.132 state jump to NORMAL
INFO  22:20:27 Updating topology for /192.168.177.135
INFO  22:20:27 Updating topology for /192.168.177.135
```

```
INFO  22:20:27 Waiting for gossip to settle before accepting client
      requests...
WARN  22:20:27 Not marking nodes down due to local pause of 11051514701 >
      5000000000
INFO  22:20:35 No gossip backlog; proceeding
```

Once the cluster is up and running, everything works the same as in a non-AWS Cassandra cluster.

Summary

A successful cluster installation depends on satisfying prerequisites. Regardless of whether you create a local cluster or one in the cloud, you can start a cluster with a minimal set of configuration properties. As with the single-node installation from the previous chapter, once you learn how to start and stop the cluster, you can configure additional configuration properties as you learn about them in the following chapters. Cassandra comes with a large number of configuration knobs and each of the following chapters will introduce a few more configuration properties.

PART II

The Data Model, Cluster Architecture, and the Cassandra Query Language

CHAPTER 4

Cassandra Data Modeling, and the Reading and Writing of Data

Data modeling in Cassandra is different from traditional data modeling in a relational database in many ways. This chapter introduces you to the key aspects of Cassandra data modeling, wherein the queries you anticipate running in the database have a lot to do with how you structure your data inside tables.

Data modeling involves identifying the types of data (entities) you want to store in a Cassandra database and the relationships among those data entities.

The key to modeling data in a Cassandra database is to focus on the following two things:

- Identifying the data access patterns

- The queries you're going to use

These two ideas will determine how you organize your data, as well as how you design and create your database tables.

Two things, queries and schema, determine the data organization. Queries are how you retrieve data from a database, and schema is how you arrange the data in the database tables. Cassandra's query-driven approach means that specific queries that you plan to use are the foundation of how you organize data. You gain efficiency for reads and writes when you group data together on the nodes by partition. The fewer the partitions a query must read, the faster the response of the database to the query.

© Sam R. Alapati 2018
S. R. Alapati, *Expert Apache Cassandra Administration*, https://doi.org/10.1007/978-1-4842-3126-5_4

This chapter explains in detail how Cassandra reads and writes data. Configurable consistency is a key Cassandra feature, and you'll learn all about read and write consistency, and how to configure various levels of consistency. Cassandra is a partitioned row store with tunable consistency. *Tunable consistency* means that the client applications determine the consistency of the data they request from the database.

Finally, this chapter explains the concepts of *linearizable consistency* and *lightweight transactions*, as well as how to ensure the atomicity of key operations with batch operations.

Cassandra and Relational Databases: Major Differences

Chapter 1 outlined the major differences between relational databases and Cassandra. Here, let's review in some detail the differences between how Cassandra and relational databases approach data modeling.

Data-Driven vs. Query-Driven Data Modeling

Data modeling in a relational database is driven entirely by data. You can also say that relational data modeling is table-driven. Normalization theory rules the roost, and this theory requires that you not duplicate data.

Once you normalize the data based on the tables and relationships among those tables, you write queries based on those tables and relationships. Typically, one table can serve multiple queries.

Cassandra organizes its data quite differently from how a relational database does it. Queries and not data drive Cassandra's data modeling methodology. This means that you organize your data based on the queries you expect that data to serve. You design your queries first and create your tables to satisfy those queries. You consider data duplication as quite normal, as a side effect of nesting data.

Cassandra precomputes queries at write time, thus optimizing writes, which means you get optimized reads as a free by-product. Relational databases compute queries when reading data. They use expensive operations such as JOIN and ORDER BY. There are no such operations in a Cassandra database.

Table Linkages and Referential Integrity

In a relational database, you combine data from multiple relations to answer a query. Referential integrity is important. In Cassandra, you must nest all the data required to answer a query within the same table. Referential integrity is not an issue at all.

Sort Differences

By default, a relational database returns rows of data in the order in which it wrote them to disk. You can use an ORDER BY clause to change the default sort order.

In Cassandra, you explicitly specify the sort order when you choose the clustering columns during the creation of a table.

Duplicating Data

As you'll learn shortly, the traditional relational database strategy of denormalizing data to avoid duplication isn't applicable to Cassandra. In a Cassandra database, you duplicate data where necessary to achieve more *efficient* reads. Cassandra capitalizes on the fact that storage costs are cheaper when compared to the cost of other components of the computational stack, such as CPU, memory, and network.

You can store the same data multiple times in Cassandra. Unlike in relational databases, data duplication is treated as a blessing and not a curse. Relational databases try to put a lot of information into their tables since the databases use the same tables for many types of searches. However, in Cassandra, having many tables with similar data is a good thing.

Remember that Cassandra is optimized for writes; it writes very fast to the storage system. Therefore, if you need to write the same data six times to disk, let it be so; multiple writes aren't going to hurt you. Following a strategy of duplicating data (which involves more writes) to enhance the performance of the read queries is an inherently sound strategy in Cassandra.

What Is Data Modeling?

Data modeling is both a science and an art. Data modeling is a structured process that involves the following:

- Collecting and analyzing the data requirements of an information system
- Identifying the entities that are part of the system and the relationships among them
- Identifying the data access patterns
- Organizing and structuring data in a specific way
- Designing and specifying a database schema
- Optimizing the schema using techniques such as indexing the data

When you're performing a data modeling exercise, you normally follow a process that consists of the following five key steps or stages:

- Analyze your requirements.
- Identify entities and relationships–the conceptual data model.
- Identify common queries–the application workflow.
- Specify the schema–the logical data model (design the tables).
- Optimize the schema–the physical data model (use CQL to implement the design). Optimizations include keys, partition sizes, and ordering.

In the following sections, I explain each of the five broad components of Cassandra data modeling.

Analyzing Your Requirements

The requirement part of Cassandra data modeling is simple. In most cases, you want to use Cassandra to solve the following problems:

- *Scalability*: Your data inflows are large and they're constantly growing.
- *Reliability*: You want an always available, highly reliable data store.
- *Ease of use*: You want a database that's easy to set up and manage.

Conceptual Modeling: Identifying the Entities and the Relationships Among Them

Ideally you must design your queries so they access a single table. You include all the attributes pertaining to an entity in a single table. This is different from a relational table design, where you store data about an entity in multiple tables and link the tables with foreign keys. Cassandra's one-table per query approach leads to faster performance.

As mentioned earlier, in a relational database, you start with a logical relationship model and then build out your physical tables based on those relationships. Cassandra follows a similar strategy, but places a heavy emphasis on considering the queries that you're planning to run against the tables in the database. Unlike in a relational mode, there are no joins and no referential integrity constraints in a Cassandra database.

Finally, there's a heavy emphasis on denormalization, which is the opposite of what you do in a relational database, where the focus is all on normalizing the data. Conceptual modeling uses the well-known entity-relationship model (ERM) to establish the entities and the relationships among them.

You denormalize data in Cassandra by designing and creating multiple tables. In recent releases, Cassandra has offered materialized views, which also enable you to create multiple views all based on the same underlying SSTables.

Reviewing the Queries You Want to Use

In this section, I use several examples to show you how to design your tables based on the queries that you anticipate.

Let's use DataStax's well-known Pro Cycling statistics for the table designs.

Note   Cassandra denormalizes data by repeating data among multiple tables. This is completely opposite to a relational database, which strives to minimize the duplication of data by normalizing it.

Example 1

Let's say you want to run a query to list each cyclist by their first and last names. Your logical model will look like the following:

```
cyclist_name
id
lastname
firstname
```

In this model,

- Partition key: `id`

- Clustering column: None

This table has the mandatory primary key, which is the column `id`. In this case, the `id` consists of just the partition key.

Example 2

The previous example was quite rudimentary. You could query just by ID, but you couldn't tell the types of races the cyclists participate in. Let's say you want to look for cyclists within specific race types. You'll want to create a different table with some of the same columns from example 1, but also add some new columns.

```
cyclist_race_type
race_type
id
points
lastname
```

In this model,

- Partition key: `race_type`

- Clustering column: `id`

This table helps you group all the cyclists by the type of race. The column `id`, which served as the partition key in the first example, now acts as the clustering column since you want to group the cyclists by ID within each partition.

Logical Modeling

The previous section on conceptual modeling showed how to design your tables based on expected queries. The next step in data modeling is to create a logical model, which contains the tables for satisfying each of your key queries, making sure that the tables include the entities and the relationships you've identified in your conceptual model.

It's during the logical modeling phase that you determine the primary keys for the tables, as well as the clustering columns to support the sort ordering your queries may require. Cassandra organizes its rows into tables. There's a mandatory primary key for each table. The primary key has multiple components, and the first component is the partition key. You can index other columns as well besides the primary key, and those other indexes are called *secondary indexes*.

Physical Data Modeling

During the physical modeling phase, you get down to the actual creation of the database objects such as tables and indexes. You nail down the data types, including any user-defined types you many need. You also determine the keyspaces you need and the partitioning and replication strategies.

In addition to secondary indexes, you must also think about any materialized views that you may need during the physical data modeling phase.

It's during this stage that you perform sizing calculations to figure out the space requirements for storing the data.

You perform several *optimizations* in the *physical* data modeling stage, including the specifying of partition sizes and ordering.

Once you have reviewed and refined the physical model, you implement the database schema in CQL by executing DDL commands such as `CREATE KEYSPACE,` `CREATE TYPE,` and `CREATE TABLE.`

Cassandra Data Modeling Rules

It's natural for developers and architects moving over to Cassandra from a relational database background to bring their traditional data modeling thinking with them. Big mistake! Many of the well-known relational database modeling principles or rules don't apply to Cassandra, and there are several new rules you ought to learn to truly benefit from using a Cassandra database. While CQL is indeed similar to SQL, avoid the temptation to create traditional data models that you're likely to be comfortable with.

To get maximum performance from a Cassandra database, you'll do well to follow certain well-established data modeling rules. Following these simple rules lets you start out with superior performance and enables you to maintain the high performance as you scale up by adding more and more nodes to your cluster.

The Two Basic Rules

There aren't many rules to Cassandra data modeling. You need to keep in mind just two rules: spreading your data across the cluster, and minimizing the number of partitions that Cassandra needs to read.

I'll elaborate on these two key rules in the following sections.

Spreading Data Evenly Across the Cluster

At an intuitive level, it makes sense that you should seek to distribute data evenly across the nodes in a Cassandra cluster. However, this isn't automatic, since Cassandra doesn't automatically move data around to balance it.

Cassandra distributes data across a cluster's nodes based on a hash of the partition key. The partition key is the first component of the primary key of a table. Thus, you need to pick a good primary key to ensure that data is balanced evenly across all nodes.

Minimizing the Number of Partitions to Be Read

In a Cassandra table, a set of rows that share a partition key is called a *partition*. Ideally, you must store the data by grouping it together by partitions on the nodes. The fewer partitions a query needs to get it data, the faster you will get the results back.

Each of a table's partitions can live on a separate node. When you issue a query, the query coordinator may issue separate commands to multiple nodes, where each of the partitions may be located. This of course means more overhead and introduces additional latency to the execution of queries. You should seek to read a query's data from as few partitions as possible.

Even if multiple partitions are stored on the same node, because of the way Cassandra stores rows in a table, it's cheaper to read the data from a single partition than from multiple partitions.

Modeling Around Queries and Not Around Relations

The way to satisfy the two basic rules, especially the minimizing of the number of partitions, is by modeling your database around your queries. Unlike in a relational database, where you model around the relations among entities, you model based on the queries you expect your database to support.

When designing a data model, always start with the queries. You need to think in terms of how the users are going to want to view the data and how they'll search through the data.

What the users are going to search for should be the primary key of the table, and the information they want to view should be your columns. That's all there's to it. You don't need to worry about all the normal forms and relationships among the data, etc.

To model around your queries, you need to do two things:

- Find out the queries the database must support.

- Create appropriate tables.

Determining the Queries

There's no single data model that serves all query cases. If you change the query requirements ever so slightly, you'll need to modify your data model. When determining the queries you want a Cassandra database to support, think of the following types of requirements in a query:

- A query that requires unique values only in the result set

- A query that wants to filter the results based on specific criteria

- A query that wants to order the results

- A query that seeks to group the results

Creating Appropriate Tables

In relational databases, the tables in most cases are just a repository for specific data such as customer data or sales data. You create the tables based on relations among the entities. A table you create in this manner services several types of queries that seek the data stored in that table. Not so in Cassandra, where your goal when creating a table is to satisfy a query by reading a single partition.

The strategy of satisfying a query by reading a single partition means that each of your queries uses a table for itself. If you have multiple queries that you need to support, you must create multiple tables, since it's unlikely for a single table to efficiently service many distinct types of queries, as is the case in a relational database.

The key to understanding the strategy here is to realize that your goal in creating a table isn't merely so it can serve as a repository for data about specific attributes of an entity. Rather, the table is a prebuilt source of answers to a common query that you must support. To optimize reads, you must create custom tables that answer that query fast.

Performance Limitations of Cassandra

As good and powerful Cassandra is, it does suffer from some well-known limitations during writes as well as during reads. I summarize the main performance drawbacks in this section.

Write Limitations

Cassandra offers a very fast write throughput, but there are a couple of key compromises that enable it to do so, as explained in the following sections.

No Support for Traditional Transactions

Unlike in a relational database, there aren't any rollback mechanisms in Cassandra. Nor do you have the traditional locking mechanisms that are the backbone of relational database transactions. Cassandra does support *lightweight transactions* as explained in Chapter 5, but these transactions are expensive.

Overhead for Mutations and Deletes

As you know by now, Cassandra stores its data in SSTables on disk. SSTables are immutable data structures. When you update data, Cassandra spreads the data across several SSTables. When you delete data, Cassandra creates *tombstones* (markers to denote data that's to be deleted) to ensure that it deletes the data correctly across the cluster. A tombstone will suppress older data until the database can run a compaction, which will remove the data for good.

Both the spreading of data across the SSTables during updates and the creation of tombstones during deletes means a higher overhead during read operations. This leads to pressure to compact the SSTables by cleaning them up.

Read Limitations

Cassandra suffers from a few known performance limitations during read operations, as I explain in the following sections.

No Support for Joins

As mentioned in Chapter 1, you can't join data from multiple tables into a single query in Cassandra. There are no foreign keys to facilitate table joins, as is true of relational databases.

Instead of joining tables, you de-normalize your data, thus duplicating it based on the expected queries. Alternatively, you can use another reporting technology such as Apache Spark to perform the joins.

Indexes Work Differently

Cassandra performs its searches via the primary key of a table, which is unique and helps identify a row very fast. Secondary indexes, employed by relational databases to speed up queries, however, are a different story altogether, and can negatively impact performance if you don't use them for the limited use-cases where they are fine.

Only Eventual Consistency

The key principle behind Cassandra's data model is tunable consistency, where the client applications determine the consistency of the data they require.

Even though Cassandra automatically replicates data across the cluster, there's an inherent latency in replicating the data, and you're bound by the principle of eventual consistency. Eventual consistency, also called optimistic replication, achieves high availability in distributed computing architectures, and informally guarantees that in the absence of newer updates to a data item, eventually accessing that data item will return its last updated value.

As explained in Chapter 1, eventually consistent services support the BASE (Basically Available, Soft, Eventual Consistency) semantics, as opposed to the ACID (Atomicity, Consistency, Isolation, and Durability) guarantees offered by relational databases.

However, you can use *quorum reads and writes* if necessary to minimize any adverse effects of eventual consistency. I explain consistency and quorum read/writes in detail later in this chapter in the "Handling Consistency" section.

The Concept of Eventual Consistency

Consistency is the requirement that a read always return the latest written data. All clients will read the same value for an element of data even when the database is simultaneously updating data.

A database such as Cassandra must make tradeoffs among data consistency, availability, and partition tolerance, which are three tenets of the CAP theorem enunciated by Brewer. Availability refers to the fact that all clients are able to access the data so they can read and write data. Partition tolerance is where a database can be split into multiple machines and be able keep functioning even during network segmentation breaks. Since network issues make temporary partitions unavoidable, you really need to choose between availability and consistency in real life.

In the real world, instead of a single concept of consistency, there are several degrees of consistency. The most stringent consistency model is strict consistency, which requires that every read must return the last written value. In a distributed system, strict consistency is hard to enforce. For example, the database will need to perform all update (insert/delete/update) operations in a synchronous fashion, using locks to prevent access to the replicas that haven't been modified yet. This of course blocks users, and if there's a failure of any kind, such as network or server failures before the update operation completes, the data will become unavailable.

Strict consistency simply won't allow you to peek at the data until the database is sure that you're looking at the most recently updated values. The database would rather become unavailable than show you inconsistent values.

Eventual consistency requires that all updates must be present on all the copies (replicas) of data in a distributed database, but allows for the fact that the process can take a bit of time. That is, although replicas may differ in value immediately following a change, over a period of time (*eventually*) all replicas will become consistent.

Cassandra implements *tuneable consistency*, where you balance the consistency level against the replication factor. The higher the replication factor, the worse the performance, but you'll gain consistency. The consistency level you choose tells the database how many replicas must acknowledge a successful write or respond to a read query for that write or read to be deemed successful.

You can specify consistency levels for both reads and writes. Higher consistency levels require more nodes to respond to a read or write operation, which means that the data is more reliable, since multiple replicas are showing identical values.

Fast writes are the goal here: although you can set the consistency level the same as the replication factor, the stronger consistency you achieve is at the cost of performance. The consistency level is usually set to a value lower than the replica factor, so updates are deemed successful even if some of the nodes are unavailable. Cassandra will keep updating data even during these partial failures.

Consistency Conflict Resolution

Since eventual consistency only guarantees that reads will eventually return the same value and doesn't make any safety guarantees, it can return a value before it converges.

To ensure replica convergence, an eventually consistency system must reconcile the differences among multiple versions of the same data items, by following this two-step procedure:

- *Anti-entropy*, which involves exchanging versions of data between the nodes

- *Reconciliation*, which involves choosing an appropriate final state of the data when concurrent updates change that data

There are several approaches to reconciling concurrent writes, such as the "last writer wins" strategy, user-specified conflict handlers, and so on. The database normally uses timestamps to detect the concurrency among the updates.

Repairing Data

Reconciliation of disparate concurrent writes must happen sometime before the next read occurs, and the database can schedule it at different points, such as the following:

- *Read repair*: The reconciliation or correction happens when a read finds an inconsistency. Obviously, this will affect the speed of the read operations.

- *Write repair*: The reconciliation or correction occurs during write operations. Any inconsistencies found during writes will slow down the write operations.

- *Asynchronous repair*: The corrections aren't part of a read or write operation.

When data is read, a read repair can resolve any data inconsistencies. When data isn't read, you need to use either the hinted handoff or anti-entropy mechanism to resolve any data inconsistences. Chapter 5 explains the various types of repairs.

How Cassandra Writes Data

Cassandra uses a multi-stage write path, which has the following stages:

- First, it logs the writes in the commit log.

- Next, it writes the data to the memtable in memory.

- Finally, it flushes the data from the memtable to permanent storage on disk in SSTables.

In the following sections, I elaborate on the three main stages of the write path.

Writing to the Commit Log to Protect the Changes

The first thing Cassandra does when writing data is to append the writes to the commit log on disk. The idea here is to protect the writes from things such as a power failure on the nodes. Thus, the commit log is a crucial component of Cassandra's durability guarantee.

The database considers a write to be successful only after it writes the data to the commit log. If a database crashes right after the database writes to the commit log, no problem. Once the database comes up, the first thing it does is replay the commit log to recover all transactions that are in there but weren't written to the SSTables. A commit log serves just one purpose: to hold committed data until it's permanently written to disk in the SSTables.

By default, the database stores the commit log in the /var/lib/cassandra/ commitlog directory, but you can specify a custom location by setting the commitlog_ directory property. Ideally, you must store the commit log on a physical device that's separate from the devices where you store the data file directories.

The database flushes memtables to disk when the space configured for the commit log reaches its limit. The database flushes the oldest commit log segments first and removes the log segments from the commit log.

When you start up the database, it replays the commit log. You can reduce the replay time by making sure the commit log isn't too big. Of course, if you configure too small a commit log, the database will be flushing data to disk quite often for active tables.

You manage the commit log by setting the following properties in the `cassandra.yaml` file:

- `commitlog_total_space_in_mb`: Configures the total space for the commit log. The default value is 8192MB for 64-bit JVMs.

- `commitlog_compression`: By default, the database doesn't compress the commit log, but you can make it compress the commit log by setting either LZ4, Snappy, or Deflate as the value for this parameter.

The database archives the commit log at the following times:

- When the node starts up

- When it writes a commit log to disk

- At a point in time that you specify

You configure commit log archiving in the `commitlog_archiving.properties` file, which is located in

- `/etc/cassandra/commitlog_archiving.properties` (package installations)

- `install_location/conf/commitlog_archiving.properties` (tarball installations)

You can archive a commit log segment with the `archive_command` command:

```
$ archive_command=/bin/ln %path /backup/%name
```

In this example, `path` refers to the path of the commit log you want to archive and name refers to the name of the commit log.

To restore an archived commit log, execute the `restore_command` command:

```
$ restore_command-cp -f %from %to
```

The `from` parameter specifies the path of the archived commit log segment in the restore directories. The parameter `to` refers to the name of the current commit log directory.

You can set the location of the restore directory by specifying the path as follows in the `cassandra.yaml` file:

```
restore_directories=<restore_directory_location>
```

Writing to the Memtable for Durability

Cassandra is designed to consume a large amount of data as fast as it can. To achieve this goal, Cassandra writes all new data first to a commit log on disk to ensure its safety, as explained earlier. Writes to the commit log are durable and survive incidents such as a power failure.

Once the database writes the change to the commit log, it writes that data also to the memtable, which lives only in memory. To provide durability, Cassandra considers the write successful once it completes writing the new data to both the commit log and a memtable.

It's important to understand that a commit log serves only to support the durability of data in case of an unexpected incident that may otherwise cause you to lose data. The memtables, on the other hand, are the mechanism that the database employs to permanently write data to the SSTables on disk.

Configuring the Flushing of Data from the Memtable

Each memtable is dedicated to storing data for a specific table, and there may be multiple memtables per table. One of these is the memtable the database currently writes data to, and the rest, memtables that are full and are waiting for the database to flush them to disk. Once a memtable reaches its memory limit (that is, once the memtable is full), Cassandra writes all the memtable's data to disk in an SSTable, which is a file. Following this, it creates a new memtable.

Cassandra flushes the memtables to disk when either the commit log space threshold or the memtable cleanup threshold is reached.

Configuring the Commit Log Space Threshold

You configure the total space used for commit logs with the `commitlog_total_space_in_mb` property. You configure the commit log on a per node basis. When the total space occupied by all commit logs in the database crosses the value you set for this property, the database flushes memtables to disk for the oldest commitlog segments.

The default value of the `commitlog_total_space_in_mb` property is 8192MB. Setting a limit for the total space that the commit log can occupy will prevent tables that you rarely update from holding onto the commit log segments forever.

If you set too low a value for the `commitlog_total_space_in_mb` property, all tables will experience more frequent flushing activity.

Note   You configure a commit log on a per-node basis, and all the tables in the database share this log. The database maintains the memtables and SSTables on a per-table basis.

Configuring the Memtable Cleanup Threshold

You can also configure the `memtable_cleanup_threshold` property to set the total amount of memory that all memtables together can use before the database flushes the largest memtable.

The `memtable_cleanup_threshold` property is now deprecated, so I won't discuss this property further.

Both the size of the data in your database and the nature of the write load will determine how you set the memtable thresholds. If there are many writes, or if the writes include a high volume of updates of small chunks of data, increasing the memtable thresholds is in order.

Manually Flushing the Memtables

You can manually flush the memtables. A recent flush of the memtables means that the database will need to replay a lower amount of commit logs, should the node restart for any reason.

> **Tip** DataStax recommends that you flush the memtable before stopping a node and starting it up again.

When a node restarts, all data in the memtable is gone. You can replay the commit log to restore the writes that were in the memtable when the node was stopped. This is so because the commit log rebuilds the memtables when you replay the log.

You can perform a manual flush of a memtable with either the `nodetool flush` or the `nodetool drain` command.

The `nodetool flush` command enables you to flush one or more tables from the memtables in memory to SSTables on disk. The command has the following syntax:

```
$ nodetool <options> flush -- <keyspace> ( <table> ... )
```

You must specify the name of the keyspace and one or more tables, separated by a space.

The `nodetool drain` command drains the entire node; that is, it flushes all memtables on this node to SSTables.

```
$ nodetool <options> drain
```

When you run this command, the database stops listening for any connections, including those from other nodes. You must restart the node after running this command.

> **Note** Use the `nodetool flush` command to flush memtables to disk. Run the `nodetool drain` command only for special occasions, such as when you're upgrading a node to an updated version of Cassandra.

The commit log is comprised of segments. When the database flushes all the data in a segment to disk, it purges the segment. Alternatively, when the commit log hits the max size that you've configured with the `commitlog_totalspace_in_mb` property, the database purges the oldest segments from the commit log, flushing the segments to disk first.

Writing to the SSTables for Durable Storage

The database flushes memtables to disk, to an SSTable. An SSTable (Sorted String Table) is immutable, meaning that once it's written, the database can't write to it ever again. If you modify the data in an SSTable, that data is written as an UPSERT (UPDATE + INSERT), and Cassandra automatically removes the previous data.

Because SSTables are immutable, and Cassandra writes to them in batches when the corresponding memtables get full, so the database doesn't have to perform random seeks. Instead, it performs only sequential I/O in large batches, which ensures a high write throughput.

In a traditional B-tree based database, the database must perform a read during a write operation to check the indexes to see where the current data is. Cassandra doesn't have to do these reads, thus keeping insert performance very high.

When Cassandra reads data, it reads both the SSTables and the memtables associated with that table since the memtables may not be flushed to disk yet and thus may contain data that's not present in the SSTables.

Format of an SSTable Data File

An SSTable consists of files on disk, and a partition usually occupies multiple SSTable files. Here's the structure of an OS-level data file that belongs to an SSTable:

`/data/ks1/cf1-5be396077b811e3a3ab9dc4b9ac088d/la-1-big-Data.db`

In this example,

- `ks1` represents the keyspace that this SSTable belongs to.

- The hexadecimal string `5be396077b811e3a3ab9dc4b9ac088d` represents a unique table ID.

- `la-1-big-Data.db` is the name of the data file.

Structures Inside an SSTable

When Cassandra writes an SSTable to disk by flushing data from the memtables, it creates a file for the data in the SSTable. Additionally, it creates several data structures along with the datafiles. These data structures are each represented by a separate file in the same directory where the database stores the data files.

The files relating to an SSTable are located in the $CASSANDRA_HOME/data/data directory. In this main directory, there'll be a directory for each keyspace, under which you'll find a subdirectory for each table. For example, for the cyclist_name table in the cycling keyspace, the following is where you'll find the files for the cycling.cyclist_ name table (the directory for the table consists of the table name along with an UUID, which helps distinguish between multiple schema versions since a table's schema can go through modifications over time):

```
$ubuntu:/cassandra/apache-cassandra-3.10/data/data/cycling/cyclist_name-
43138460591411e7b1387bff0507f153# ls -altr
total 48
drwxr-xr-x 2 cassandra cassandra  4096 Jun 24 12:35 backups
drwxr-xr-x 5 cassandra cassandra  4096 Jun 24 12:36 ..
-rw-r--r-- 1 cassandra cassandra    16 Jun 25 08:39 mc-1-big-Filter.db
-rw-r--r-- 1 cassandra cassandra    20 Jun 25 08:39 mc-1-big-Index.db
-rw-r--r-- 1 cassandra cassandra    92 Jun 25 08:39 mc-1-big-Summary.db
-rw-r--r-- 1 cassandra cassandra    57 Jun 25 08:39 mc-1-big-Data.db
-rw-r--r-- 1 cassandra cassandra    10 Jun 25 08:39 mc-1-big-Digest.crc32
-rw-r--r-- 1 cassandra cassandra    43 Jun 25 08:39 mc-1-big-
CompressionInfo.db
-rw-r--r-- 1 cassandra cassandra  4660 Jun 25 08:39 mc-1-big-Statistics.db
-rw-r--r-- 1 cassandra cassandra    92 Jun 25 08:39 mc-1-big-TOC.txt
drwxr-xr-x 3 cassandra cassandra  4096 Jun 25 08:39 .
ubuntu:/cassandra/apache-cassandra-3.10/data/data/cycling/cyclist_name-
43138460591411e7b1387bff0507f153#
```

I summarize the main data structures here:

- Data (Data.db): This is the main SSTable data file, and the only file stored by Cassandra's backups.

- Primary Index (Index.db): An index of the row keys with pointers to the key locations on the data file.

- Statistics (Statistics.db): Contains the statistical metadata about the SSTable's data.

- Compression Information (Compressioninfo.db): A file that contains information regarding the compression of the SSTable.

- Secondary Index (`SI_.*.db`): The built-in secondary index. There may be more than one of these per SSTable.

- SSTable Index Summary (`Summary.db`): A sample of the partition index that the database stores in memory.

- Bloom Filter (`Filter.db`): This file contains the Bloom filter for this table. The Bloom filter is a memory structure that helps the database quickly check if row data is in the memtable, before accessing the SSTables. Bloom filters for an SSTable enhance read performance.

You can compress SSTables to save storage. I explain SSTable compression in Chapter 11.

Caching SSTable Data

Cassandra provides three types of caching for SSTable data:

- A *key cache* stores a map of partition keys to row index entries, which allows fast access to SSTables. In simpler terms, the key cache stores partition indexes for tables.

- A *row cache* caches frequently accessed rows to speed up access to those rows.

- A *counter cache* improves performance for the counters that you'll need to access frequently.

Key and counter caches are enabled by default. Cassandra stores cached data to disk so it can quickly pull it into memory when you restart the database. I explain caching in detail in Chapter 11.

The Write Request Flow

Cassandra processes data in multiple stages on the write path, which starts with the initialization of the write request by a client and ends with the database storing the data on disk. It immediately logs a write, and at the end of the process, writes the data to disk. Here are the stages of a write:

- Log data in the commit log.

- Write the data to the memtable.

- Flush the data from the memtable.

- Store the data in SSTables on disk.

Figure 4-1 shows the Cassandra write request flow. The figure shows a Cassandra cluster with twelve nodes. There's three-way replication (that is, the replication level is 3), and the clients use the QUORUM consistency level.

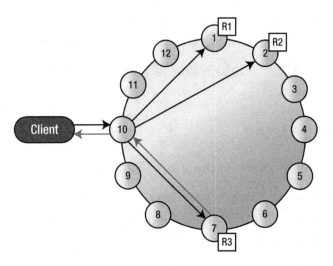

Figure 4-1. *The Cassandra write request flow*

Note   Cassandra writes sequentially and it doesn't need any reads or seeks to write values since all writes are just appends. Therefore, Cassandra can write data very fast.

Under the QUORUM consistency level, Cassandra deems a write operation a success if two of three nodes acknowledge success of the write operation. The replica nodes send the acknowledgements to the coordinator node, which is the node to which the clients connect.

The following is a step-by-step explanation of the Cassandra write request flow shown in Figure 4-1.

1. A Cassandra client sends a write request to store a specific key. The first thing to understand is that a node can play the role of a replica, a coordinator, or both. If the node is the right partition to store the key, and if the node maps to the data, the node acts as the replica. If the node happens to be the node the client connects to, then it acts as the coordinator. If both things are true, the node plays both roles: coordinator and replica.

2. The coordinator figures out the right replica nodes that should store the key and forwards the client's request to all those nodes. The number of replicas depends on the replication factor. If all nodes are not up, the nodes that are down will have inconsistent data after the writes. The down nodes, once they come back, will use one of Cassandra's data repair mechanisms, such as hinted handoff, read repair, or anti-entropy repair, to become consistent.

3. The nodes that receive the keys sequentially write the data, along with the metadata for creating the data, in the commit log (locally).

Tip   A hint counts as a successful write only under the consistency level ANY.

---

4. The nodes then write the keys and their data locally to the memtable, which lives in memory. At this point, the nodes consider the writes a success.

5. Once the write succeeds (or fails), the replica nodes send the operation's success/failure status to the coordinator node.

6. In this case, we're assuming a consistency level of QUORUM and a replication factor of 3. The coordinator will therefore respond with a success message to the client once two of the three nodes respond with a success message.

> **Note**   You'll often come across the term *coordinator node*. The coordinator node is the node where you initiate a cqlsh session. The coordinator node can be a remote host if you connect to one.

There are several important Cassandra internal operations that occur around this time, such as the following:

- How Cassandra uses hints during the write operations

- How Cassandra determines when it should flush memtables to disk

- How Cassandra uses the commit log

- The role that Bloom filters play

- The role of index files

The following sections explain these behind-the-scenes internal operations or processes.

The Role of Hints During the Write Process

If one or more of the nodes fails to send an acknowledgement of success or failure, the coordinator stores a hint locally, so it can resend the write operation to the node(s) when they come back up again.

You can control how Cassandra manages hints with the help of the following two configuration properties (in the `cassandra.yaml` file):

- `gc_grace_seconds`: Configures the time-to-live (TTL) period for a hint so that the database won't replay the hint after this duration. Setting this parameter to 0 disables hints.

- `max_hint_window_in_ms`: This property determines the period for which hints will be recorded. The default value is three hours.

Memtables and How They're Flushed to Disk

During client write operations, a background thread monitors all the memtables in the database. Once any of the following conditions is satisfied, the background thread replaces the current memtable with a new memtable, and marks the older memtable for flushing to disk (to an SSTable):

- The node hits its global memory threshold.

- The commit log is full.

- The table level interval is reached.

A different background thread (or a set of threads) flushes the memtables that the first background thread has marked for flushing.

The Commit Log

Once Cassandra flushes a memtable to disk, all the entries stored in the commit log for the keys in that memtable become useless. Cassandra marks all the commit log segments belonging to the flushed memtable so that it can use them again for storing data from other memtables.

The database stores the commit logs in the file system as binary files in the $CASSANDRA_HOME/data/commitlog directory. Here's an example:

```
$ ls -altr
total 9332
drwxr-xr-x 6 cassandra cassandra      4096 Apr  3 10:56 ..
-rw-r--r-- 1 cassandra cassandra  33554432 Jun 24 12:35
CommitLog-6-1498331957174.log
drwxr-xr-x 2 cassandra cassandra      4096 Jun 24 12:35 .
-rw-r--r-- 1 cassandra cassandra  33554432 Jun 25 09:55
CommitLog-6-1498331957173.log
$
```

The number 6 in the commit log name shows the number version of the commit log format. For the Cassandra 3.0 release, the version number is 6.

Bloom Filters and Index Files

Once Cassandra flushes a memtable to disk, it creates two additional data structures: a Bloom filter and an index file.

Bloom Filters

A Bloom filter is a probabilistic data structure for testing set membership that you can tune for false positives. It never produces a false negative. A Bloom filter is an off-heap structure. It uses a fast algorithm to test whether an element is a member of a set.

The database stores the Bloom filters in memory and uses them to reduce disk accesses when looking up keys. When you perform a query, the database first checks the Bloom filter before accessing data on disk. If the Bloom filter shows that the element is in the set, the database accesses the disk to ascertain it. Since memory access is much faster than disk access, you can view a Bloom filter as a type of cache to speed up access.

When a Bloom filter reports that a key isn't present in an SSTable, then the key isn't present. However, if the filter reasons that the key is present, it may be wrong at times; the key might or might not be actually present.

Bloom filters serve to enhance the scalability of read requests. They do this by keeping the database from performing the additional disk reads for reading the SSTable by indicating if a key isn't present in a SSTable.

You can reduce the number of false positives by increasing the size of the Bloom filters. You'll use additional memory but you'll be making the filters more accurate. You can configure the chance of a false positive at the table level, either when creating the table or later. You configure the `bloom_filter_fp_chance` property at the table level to specify the percentage of false positives a Bloom filter reports.

Index Files

An index file stores the offset of keys into the main data file, which is the SSTable. By default, the database stores a part of the index file in memory. The index file stores the offset for every 128th key in the SSTable, and you can configure this value.

As with Bloom filters, index files enhance read scalability. They do this by providing the random position from an SSTable where you can sequentially scan to get the data. Otherwise, you'll be forced to scan the entire SSTable to retrieve the data.

Compacting SSTable Data

Over time, the database can end up with multiple versions of a row in different SSTables, since SSTables are immutable and the database doesn't overwrite existing data. Instead, inserts and updates result in new SSTables.

To keep the multiple versions of the SSTables from overwhelming it, the database merges SSTables on a regular basis to get rid of the older versions of the data. This process is called *compaction*. By default, the database automatically performs minor compactions.

During a compaction operation, Cassandra merges the data in SSTables, including the keys and columns. It also gets rid of all the expired tombstones and creates a new index. The outcome of the compaction is a single new SSTable. Compaction may result in a temporary rise in I/O usage and the size of data on disk, but it ends up saving you space since it merges several large datafiles that store the SSTable data.

You can configure a compaction strategy to tell the database which algorithm it must use when it merges multiple SSTables during a compaction. For example, the default compaction strategy, size-tiered compaction, merges similar sized SSTables to create one large SSTable.

Once the database creates the new SSTable by merging multiple SSTables, it marks the older tables for deletion. This means you'll see a higher space usage during a compaction operation. The database gets rid of the tables it marks for deletion during a restart, or by using a reference counting mechanism.

After a compaction, the database will remove the original SSTables and thus gets rid of the outdated rows. However, compaction is on a per-node basis, so although one node performs a compaction, outdated versions of rows may still exist on other nodes. To avoid returning the outdated rows, the database gets multiple versions of a row in response to a read request from a client. It then returns the version with the latest timestamp to the client, a process called "last-write-wins."

Cassandra offers multiple compaction algorithms, and different strategies are best for write heavy and read heavy tables. I discuss compaction strategies in detail in Chapter 11.

Reading Data

I want to explain the read request flow in a Cassandra database, but before I do that, let me explain the basic architecture of a Cassandra database.

Since there's no primary or master node in a Casandra cluster, any node that contains the rows that can satisfy queries about that row can do so. Remember that Cassandra replicates each row in a table across multiple nodes, with the number of replicas depending on the replication factor.

The *gossip protocol* that Cassandra uses lets the nodes exchange information about the network topology. Using Gossip, each of the nodes learns about the cluster topology and determines where it should direct a request for a specific row.

Cassandra deems a read operation successful if enough nodes acknowledge success. How many nodes are enough depends on the replication factor and the consistency level you choose.

Figure 4-2 shows a case where there are twelve nodes in the cluster and the replication factor is 3. The QUORUM consistency level is employed during client reads. Using the QUORUM consistency level, a read operation is successful when two of the three nodes acknowledge success.

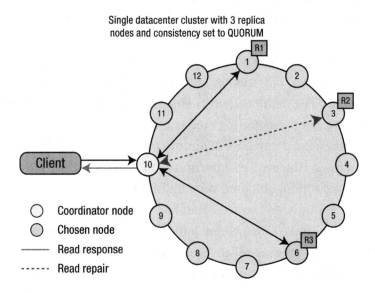

Figure 4-2. *How the gossip protocol helps nodes exchange information among themselves*

Note   Since Cassandra uses the gossip protocol and since it's a distributed database that requires it to talk to multiple nodes during each read and write operation, the network performs a heavy amount of data transfers. You're advised to set up at least a 1Gbps network bandwidth to accommodate gossip traffic as well as replication traffic.

The Cassandra Read Path

Cassandra follows a specific sequence of events before it can successfully complete a read request. The following sequence of events shows how the database serves a read request.

1. The first event is when a client sends a read request to the cluster to get data for a specific key, named K. A node can act as a coordinator, a replica, or both. If the node has the replica, it can act as the replication node. If the key isn't mapped to this node, it acts as a coordinator.

2. If the node is acting as the coordinator, it forwards the request to the nodes that might contain the key K. The coordinator uses a snitch to find which of the replica nodes is closest to itself and sends a request for full data to that node. It also sends a request to the other nodes for the digest generated with the hash of the data.

3. Cassandra forwards the request from the coordinator to the internal services of the target node(s).

4. The coordinator requests data from memtables and SSTables. Although the database can return the data faster from the memtables since they live in memory, it also checks one or more SSTables for the data. When one of the Bloom filters confirms that it has the key, Cassandra will then check the sample index in memory.

5. The database first performs a binary search on the sample index to find the starting offset into the index file. Using this offset, the database offsets into the index file and performs a sequential read to get the offset into the SSTable for the actual key.

6. Using the offset for the actual key, the database offsets into the SSTable file and returns the actual data from the SSTable by performing a lookup of the SSTable.

7. The `filter` command consolidates the multiple versions of the key data it received from the SSTable lookup and from the memtable. It consolidates the key data and returns the consolidated version to the internal services.

8. The internal services repeat this process on other nodes and return the results to the coordinator node.

9. The coordinator node compares all the digests it received from all the nodes, and if there's a conflict, reconciles the data and returns the reconciled version to the client. If necessary, the database also initiates a read repair to make the data consistent.

How Write Patterns Affect Reads

The way in which Cassandra writes its data has a bearing on the read operations. You can affect read performance by how you configure Cassandra's compaction processes. The data in the SSTables that Cassandra writes to disk is immutable. Cassandra doesn't update data. It writes new data as well as updated data to new SSTables, each with a different timestamp. The number of the SSTables may grow to be quite large over time. When responding to a query, the database may have to access many SSTables to retrieve a single row since each of the versions of a row with a unique set of columns may be stored in a different SSTable.

Cassandra periodically compacts a collection of SSTables by merging multiple SSTables and removing old data. It compacts multiple SSTables so it can make a single complete row from all the versions of each row, with the latest version of each of the columns in the row.

THE CASSANDRA STORAGE ENGINE

Cassandra uses a log-structured storage engine that avoids overwrites and employs sequential I/O to update data.

Most relational databases use a B-tree, a data structure that keeps data sorted and permits you to search, insert, delete, and sequentially access data in logarithmic time. The B-tree data structure is optimized for reading and writing large blocks of data. Cassandra uses a different data structure called the Log-Structured Merge-Tree (LSM tree) that works well for providing indexed access to files with high insert volumes. Data stores such as HBase, MongoDB, and InfluxDB also use the same LSM tree data structure.

In a large distributed database, read-before-write can lead to latencies in reads. Cassandra avoids using read-before-write for most writes. Instead, Cassandra's storage engine groups inserts and updates in memory and writes data periodically and sequentially to disk. Once Cassandra writes the data to disk, the data remains immutable; you can't overwrite it.

Write amplification (WA) is a phenomenon associated with flash memory and solid-state drives (SSDs) wherein the data physically written to the storage media is a multiple of the logical amount of data. By sequentially writing data to immutable files on disk, Cassandra avoids the write amplification phenomenon. Thus, write amplification, which is a problem on SSDs when using many other types of databases, ceases to be a problem with Cassandra.

Cassandra Transactions and the ACID Properties

Unlike traditional RDBMS transactions that support all the well-known ACID transactions, Cassandra supports the atomicity, isolation, and durability properties, but not the consistency property. Cassandra doesn't use a typical database's rollback and locking mechanisms. It offers eventual or tunable consistency wherein you determine the transaction consistency level, which involves an inherent tradeoff between the availability and the accuracy of data.

In the following sections, let's briefly review the support for the standard ACID properties in a Cassandra database.

Atomicity

Cassandra performs both writes (inserts and updates) and delete operations atomically at the partition level. Dealing with operations at the partition level means that Cassandra treats inserts, updates, and deletes of multiple rows from the same partition as a single operation.

If more than one client updates the same column simultaneously, the latest timestamp determines which is the most up-to-date update to that column.

Cassandra doesn't automatically roll back writes that are successful only on some nodes. Let's say you are using a write consistency level of QUORUM, with a replication factor of 3. This means that the coordinator nodes will need to wait for acknowledgement from two nodes. If the write succeeds on one node but fails on the other, Cassandra doesn't roll back the successful write.

Isolation

Cassandra writes and updates are fully isolated at the row level, meaning that a client that writes to a row in a partition is the only one that can see the write, until it completes the write operation.

Any updates that belong to the same partition in a batch operation are performed with full row-level isolation as well, unless that operation involves changes to multiple partitions.

Durability

Cassandra provides strong durability for writes. It provides local durability by recording all writes to both memory and to the commit log on disk before it acknowledges a successful write. If there's a server failure before the database can flush the memtable to disk, it replays the commit log to retrieve the lost writes.

The fact that Cassandra replicates data to multiple nodes bolsters the durability provided by writing to the commit log.

Specify the `commitlog_sync` property (in the `cassandra.yaml` file) to set durability to fit your need for consistency. The `commitlog_sync` property determines the method that Cassandra uses to acknowledge writes. Here are the ways in which you can configure this property:

- *Periodic* (Default value is 10000 milliseconds, which is 10 seconds): If you choose periodic, set the `commit_sync_period_in_ms` property to control how often Cassandra synchronizes the commit log to disk. Cassandra acknowledges periodic syncs immediately.

- *Batch* (Default: disabled): When you enable the batch method, Cassandra waits to acknowledge writes until they're synced to disk. The property `commitlog_sync_batch_window_in_ms` lets you specify the time for which Cassandra waits for other writes before performing a sync. The default value is 2 ms.

Handling Consistency

Cassandra enables you to configure consistency levels to balance the availability of data with the accuracy of that data. You configure consistency at a session level or for a single read or write operation. The level of consistency you choose determines the number of replicas that the database must acknowledge to the application to indicate that a read or write operation was successful.

Consistency levels determine how many nodes must respond to the coordinator node to successfully process transactions. Cassandra uses the consistency levels for handling regular (non-lightweight) transactions.

Consistency shows how synchronized all copies of a row of data are. While regular repair operations can keep the variability of data among the replicas low, the constant flow of traffic through the database can potentially lead to some replicas being out of order with the others. That is, some replicas may be inconsistent or stale.

While Cassandra does satisfy the acid, isolation, and durability properties, it's the way it satisfies the consistency property that sets it apart from a typical RDBMS.

Of the three tenets of the CAP theorem (consistency, availability, and partition tolerance), Cassandra provides the availability and partition tolerance features, and you can also configure the database so it provides the consistency and partition tolerance capabilities.

For both reads and writes, Cassandra offers *tunable consistency*, which is an extension of the principle of eventual consistency. You can configure the consistency for specific read and write operations so the data is consistent and satisfies the application requirements. This adjustability allows Cassandra to be either a consistency and partition tolerant system or a highly available and partition tolerant system. That is, it allows you to configure Cassandra to satisfy any two of the three CAP requirements.

By adjusting consistency, you're indirectly adjusting performance, since there's an indirect relationship between consistency and latency; the higher the consistency you configure, the lower the latency, and vice versa.

Note   Consistency levels determine the tradeoff between the availability of data (no latency) and the accuracy of data.

Cassandra offers both read and write consistency levels that you can configure to suit your requirements, as I explain in the following sections.

Write Consistency

You can configure a write consistency level to specify how many replicas must acknowledge a write request before the database considers the write a success. Once the number of replicas you specify report a successful write, the coordinator node that sends out the write requests informs the application that the write is completed.

How does a write operation handle a downed or unavailable node? This is where Cassandra utilizes its *hinted handoff* feature to make sure that it writes the missing data to an unavailable node once it comes back up. The database stores the missed writes as hints on the coordinator node (in the `system.hints` table) and writes the missed data to the nodes later on when they become responsive again.

You can check the default consistency level by issuing the `CONSISTENCY` command:

```
cqlsh> consistency;
Current consistency level is ONE.
cqlsh>
```

Client code can override the default consistency level by specifying the required consistency level in an individual statement:

```
statement = ...
statement.setConsistencyLevelConsistencyLevel.LOCAL_ONE);
```

You can configure various write consistency levels. In the following sections, I describe the write consistency levels from the strongest (`ALL`) to the weakest (`ANY`).

ALL (Strong Consistency)

The `ALL` consistency level provides the highest level of consistency, but the tradeoff is that it also offers the lowest availability compared to the other consistency levels. Under the `ALL` consistency level, for a write to a partition to be successful, the database must perform the write to the commit log and the memtable on *all the replica nodes* in the cluster. This means that the number of nodes that must reply should be the same as the replication factor you configure. The database marks the write operation a failure if even one of the replicas doesn't respond.

Quorum-Related Levels (Strong Consistency)

Cassandra offers several levels that are *quorum-based*, all of which offer a strong level of consistency. The following are the different quorum-based consistency levels:

- `EACH_QUORUM`: Provides strong consistency by requiring a write to be written to the commit log and the memtable on a quorum of the replica nodes in *each of a cluster's datacenters*. You specify the `EACH_QUORUM` level of consistency in a multiple datacenter cluster when you want to ensure there's strict consistency at the same level in all datacenters. Setting this consistency level means that a read will fail if a datacenter is down, making it impossible to reach the quorum in that data center.

- QUORUM: Provides strong consistency. Under this consistency level, Cassandra deems a write successful if it's written to both the commit log and the memtable on a quorum of nodes. Unlike in the EACH_QUORUM consistency level, the nodes can range across multiple datacenters in the cluster. You can specify this consistency level either in a single or a multiple datacenter cluster. This level helps you maintain strong consistency by considering the entire cluster as a unit instead of a datacenter.

- LOCAL_QUORUM: Provides strong consistency. This consistency level requires the database to complete the write to both the commit log/memtable on a quorum of local nodes, that is, nodes that belong to the same datacenter as the coordinator.

  You can use LOCAL_QUORUM consistency only in a multiple datacenter cluster with a replica placement strategy such as NetworkTopologyStrategy and an appropriately configured snitch.

Note   When you perform a write operation with the QUORUM consistency level, you'll always get the correct data, even when one of the replicas dies.

One, Two, Three, and LOCAL_ONE Consistency Levels (Weak Consistency)

The ONE consistency level isn't strict. It requires that the database perform a write to the commit log and memtable of at least one replica node. The TWO and THREE consistency levels are similar but stricter since they require the write to be made to two and three replica nodes, respectively.

When you have multiple datacenters in your cluster, a consistency level of ONE is fine and is desirable as well. However, when a datacenter is offline, connections are made automatically to other online nodes in other datacenters.

The lowest level of consistency to achieve high performance and durability is consistency level ONE. Even if the single node goes down, since it already wrote the data to the commit log, it can replay the commit log when it comes back up again.

The LOCAL_ONE consistency level prevents cross-datacenter traffic by requiring a write to be acknowledged by at least one node in the local datacenter.

Note the following:

- If your consistency level is ONE and a replica crashes soon after a write, the data is lost.

- If your consistency level is ONE and the write operation times out, a subsequent read operation can return either the old or the new value. You just won't know if the data is correct.

- If your consistency level is ONE, a down node will be showing you stale data when it's revived, until the node gets the correct data or the database completes a read repair operation.

ANY

The ANY consistency level offers the highest availability, but at the cost of consistency, since this level offers the lowest consistency. This level guarantees that a write will never fail, but you pay for this guarantee with slower writes. Here's how this consistency level works:

- The database must perform the write on at least one node in the cluster.

- If none of the nodes for a partition key are available, the write succeeds only after one of the nodes comes back up, and, using the hinted handoff feature, catches up with the lost writes. This level of consistency therefore allows a hint to count towards a successful write.

The consistency level ANY is meant for applications that want the database to accept writes even when all of the replica nodes are down, that is, even when the database can't satisfy the consistency level ONE. The consistency level ANY guarantees that a write is durable and will be readable by a database once the target replica becomes available and it receives the replayed hints.

Calculating the Quorum

The concept of a quorum plays a significant role in configuring several write and read consistency levels. A quorum refers to the number of nodes that need to respond for Cassandra to consider a write or read operation successful.

When computing the quorum in a single datacenter cluster, you need to consider the replication factor for just one datacenter. In a cluster with multiple datacenters, you consider the replication factor in all datacenters, which need not be the same. The more datacenters in a cluster, the higher the quorum, since more replica nodes need to respond for Cassandra to consider a read/write operation successful.

Quorum-based consistency levels require the writing of data to at least the number of nodes that constitute a quorum.

You can calculate the quorum with the help of the following formula:

quorum = (sum_of_replication_factors/2) + 1,

where the value of the sum_of_replication_factors is the sum of all the replication_factor settings for all datacenters within the cluster.

For example, of the replication factor is 3, in a single datacenter cluster, the quorum is 3/2 plus 1 which gives you 2.5, but you round it down to a whole number, which makes it 2.

A quorum of 2 means that the cluster with a single datacenter can tolerate one replica being down. If the quorum is 4, the cluster can tolerate two nodes down in a six-node cluster.

It's important to understand that LOCAL_QUORUM, which is similar to QUORUM, considers the replication factor of the datacenter that contains the coordinator node. Cassandra will consider only local replica nodes and ignores other datacenters while computing the quorum.

How Consistency Works in Practice

The replication factor and the read and write consistency levels that you configure will together determine how reliable your read and write operations are.

For example, if you satisfy the following, the database can guarantee eventual consistency:

$$R + W <= N$$

where

- R is the consistency level of read operations.

- W is the consistency level of write operations.

- N is the replication factor.

Let's say you set the replication level to 3. A read operation that uses the QUORUM consistency level and a write operation using the ONE consistency level will then ensure eventual consistency. This is so because the QUORUM consistency level needs to read two replicas and the ONE consistency level for writes needs to write to one replica, making the total three replicas, the same number as the replication factor.

If you need a stronger consistency, you can guarantee it by making sure that the consistency levels for reads and writes together are at least four when your replication level is 3. This means that a read operation can use two replicas to verify data and the write operations will use two replicas as well to verify the writes, yielding a strong consistency. In other words, you can guarantee strong consistency by making sure the following condition is true:

$$R + W > N$$

If speed is of the essence, you can achieve it while still providing strong consistency by lowering the write consistency level to 1 but raising the read consistency level to 3. You're going to write data faster but your read operations will have additional latency as a result.

Read Consistency

As with write consistency, there are multiple read consistency levels you can configure. And, as in the case of write consistency, you face a tradeoff between consistency and availability when choosing from various consistency levels. For a read operation, the consistency level you configure determines how many replica nodes must respond successfully to a read request before the database returns the results. Eventual consistency means other work keeps going on in the background while the database is waiting on the information from the nodes.

Since Cassandra is a distributed database, the most recent value of data that you query isn't necessarily present on every node of the cluster at any given time. Client applications configure consistency levels for their requests to manage the tradeoff between response time and data accuracy.

A higher consistency level requires more nodes to respond to a query, thus providing more of an assurance that the data in each replica are identical. If two replicas return data with different timestamps, Cassandra returns the data with the more recent timestamp to the client. It then initiates a read repair to update the values on the replica with the outdated values so it has the latest value as well, thus ensuring that it's consistent.

Briefly, you can configure the following read consistency levels, each with a different number of replicas that participate in satisfying read requests:

- ONE: Returns data from the nearest replica

- QUORUM: Returns the most up-to-date data from a majority of the replicas

- ALL: Returns the most up-to-date data from all the replicas in a cluster

The following sections summarize the availability and consistency tradeoffs for each of the read consistency levels.

Note   Be sure to understand the differences and the relationship between the replication factor and the consistency levels. You set the replication factor at the keyspace level and the clients specify the consistency levels per query. The consistency level is based on the replication factor, since the replication factor determines the number of replicas that store data during a write. And the consistency level determines the number of nodes that should respond to indicate the success of a read or write operation.

ALL (Strict Consistency)

The ALL consistency level provides the highest consistency at the expense of availability. When you set this level, the database returns your query results only after it has heard from all of the replicas. If any of the replicas don't respond, the query fails. As with many of the consistency levels, the database performs a read repair if necessary.

When you specify the ALL consistency level, the read operation fails if any node on the cluster with that data is unresponsive at that time or fails to respond within the timeout. You configure this timeout with the rpc_timeout_in_ms property. The default timeout is 10 seconds.

Quorum-Related Read Consistency Levels (Strong Consistency)

There are two quorum-based consistency levels:

- QUORUM: Provides strong consistency by returning query results only after a quorum of replicas from all datacenters respond. After a majority of the replicas respond, the database returns the value with the latest timestamp to the client. If the database finds that some nodes have stale data, it performs a read repair on the replicas with the outdated data.

- LOCAL_QUORUM: Provides strong consistency by returning query results only after a quorum of nodes in the local (current) datacenter has replied. You can use this level only in a multiple datacenter cluster with the NetworkTopologyStrategy replica placement strategy in place and an appropriately configured snitch.

Note   A QUORUM read consistency level means that you'll always retrieve the correct data, even when one of the replicas dies.

The ONE, TWO, THREE, and LOCAL_ONE Consistency Levels (Weak Consistency)

The ONE consistency level offers the highest availability at the cost of consistency. Cassandra returns the query results immediately from the nearest replica. The tradeoff is that you have a high probability of reading stale data. The database checks this data against data in the other replicas as well, and if any of the replicas are out of date, the database initiates a read repair to make the other replicas consistent.

Note   Under the ONE consistency level, the replicas that serve the reads may not have the latest version of the data.

The TWO, THREE, and LOCAL_ONE levels of consistency are similar to the ONE setting:

- Level TWO returns the latest data from two of the nearest replicas.

- Level THREE returns the latest data from three of the nearest replicas.

- The LOCAL_ONE consistency level is similar to the consistency level ONE
  and returns the data from the nearest replica in the local datacenter.

The Serial Consistency Settings

There are two SERIAL consistency levels you can configure. The first one, SERIAL, enables you to read the latest value of a column involved in an in-flight lightweight transaction by a user. The SERIAL consistency level allows the reading of the most up-to-date data, including potentially uncommitted data.

The LOCAL_SERIAL consistency level is like the SERIAL level of consistency but is limited to the datacenter. You use this to gain linearizable consistency for lightweight transactions.

Note   You configure consistency levels to trade off availability vs. the accuracy of data.

Configuring the Consistency Level

You can configure the consistency levels at a session level or per each read or write operation.

If you're using cqlsh, specify the keyword CONSISTENCY to set the consistency level for all queries you run within your current session. Client applications set the consistency level via their drivers.

The default consistency level is ONE for a read or write operation. The following example illustrates this:

```
cqlsh> CONSISTENCY;
Current consistency level is ONE.
cqlsh>
```

Earlier, you learned about the various consistency levels you can set. Here are some examples that show how to set various consistency levels.

- Set the QUORUM level of consistency to force a majority of the nodes
  in a cluster to respond before the database considers the operation
  successful.

  ```
  cqlsh> CONSISTENCY QUORUM;
  ```

- For lightweight transactions, set the SERIAL consistency level.

  ```
  cqlsh> CONSISTENCY SERIAL;
  ```

Three Types of Read Requests

A coordinator node can send out three types of read requests to a replica node:

- *Direct Read Requests*: The coordinator node contains a single replica node.

- *Digest Request*: The coordinator first contacts a replica node, following which it sends a request to the number of nodes you specify with the consistency level property. The request sent by the coordinator checks the data on the replica nodes to ensure it's up to date. If the coordinator discovers that some nodes have stale data, it sends out a *read repair* request.

- *A Background Read Repair Request*: Cassandra performs read repair in the background to ensure that all the rows that a coordinator requests during a digest request are consistent across all the replicas that are part of the read request.

Rapid Read Protection with Speculative Retrying

Normally, during a read request, Cassandra sends out data requests to enough replicas to satisfy the consistency level. You can configure a table with the speculative_retry property to take advantage of Cassandra's ability to retry a read request with a different replica node when the original replica node is unavailable or is very slow in responding to a read request. When you configure this, Cassandra sends extra read requests to replicas, although the consistency level has been already satisfied.

Note   Rapid read protection assures that if a replica node fails or is unduly slow, the coordinator node automatically sends the read request to other nodes after a timeout interval.

When you configure speculative retrying, the coordinator node automatically retries requests with a different node after a set time elapses. Recovering from replica node failures in this manner using speculative retrying of read queries is called *rapid read protection.*

You can configure rapid read protection by setting the `speculative_retry` property. This property determines exactly when the coordinator node fires the extra read requests. Here are all the ways in which you can configure this property:

- `NONE`: This is the default value and it specifies that the coordinator not send any extra read requests except for the original read request, regardless of the latency.

- `ALWAYS`: After every read of a table, the coordinator sends extra read requests to all other replica nodes in the cluster.

- `Xpercentile` This tells the coordinator node to send out redundant read requests if it doesn't receive any responses from the replica node within a certain percentage of the value you set for the `Xpercentile` property. For example, the typical latency for a table is 60 milliseconds, and say you set the `Xpercentile` property to 80 percentile. If the replica node doesn't respond within 48 ms (80% of 60ms), Cassandra sends out redundant read requests to other replicas.

- `Nms`: Under this strategy, the coordinator node sends extra read requests if it doesn't receive any responses from a replica within N milliseconds.

Here are some examples of how to set the `speculative_retry` property for a table:

```
cqlsh> ALTER TABLE users WITH speculative_retry =  '5ms";
cqlsh> ALTER TABLE users WITH speculative_retry = '95percentaile';
```

Read Requests with Various Read Consistency Levels: Some Examples

In this section, I present several examples of read requests with varying consistency levels in a single datacenter as well as in a cluster with two datacenters. The replication factor is 3 in all cases. I first show the scenarios for a single datacenter and then for a cluster with two datacenters.

Single Datacenter

With a QUORUM read consistency level, two of the three replicas must respond before the read request is successful. If there are multiple versions of a row, the replica with most recent version will satisfy the read request. In addition, the database will initiate a read repair on the third replica if necessary (if the third replica is stale).

A read consistency level of ONE means that that database will reach out for the nearest replica for the rows it needs to satisfy the read request. Depending on whether the data diverges among the three replicas, the database may also start a read repair for the other two replicas. This depends on how you configured the read_repair_chance property for the table.

A Cluster with Two Datacenters

If you set read consistency to QUORUM and a replication factor of 3, four replicas of data must respond for a read request to succeed, with the replicas belonging to either datacenter. The database checks the other replicas for consistency, and if any replicas are out of date, starts a read repair to bring them up to date.

If you've configured the LOCAL_QUORUM level of read consistency instead, two replicas from the same datacenter as the coordinator node must respond to the read request.

For a read consistency of ONE, the database relies on the nearest replica, irrespective of the datacenter. It may also initiate a read repair based on the read_repair_chance setting you've configured for the table. For a read consistency level of LOCAL_ONE, the database contacts the closest replica in the same datacenter as the coordinator node.

Testing the Performance of Consistency Levels

As you just learned, there are many consistency levels for both read and write operations. Selecting a consistency level has a significant bearing on query and write latency, as well as on availability.

DataStax recommends that you test the performance of various consistency levels by using CQL's TRACING command. Since the TRACING command's output shows the elapsed time for each read and write operation, you can compare the performance of various consistency levels before settling on one of them.

To trace queries that use large data sets, it's a good idea to configure probabilistic tracing. Use the `nodetool settraceprobability` command to configure probabilistic tracing. Once you do this, you can query the `system_traces` keyspace, as shown here:

```
cqlsh> SELECT * FROM system_traces.events;
```

Just a disclaimer here regarding the performance hit of probabilistic tracing. Experience shows that the tracing could significantly affect database performance. Therefore, use this feature with caution.

When you turn tracing on, if you have a replication factor of 3, here's what the three consistency levels would mean:

- ONE: Processes responses from one of the three replicas.

- TWO: Processes requests from a majority of the three replicas (that is, two replicas).

- ALL: Processes responses from all three of the replicas.

To trace the queries, all you need to do is to run the query you want to use for testing the differences among the consistency levels, after turning tracing on:

```
cqlsh> TRACING on;
cqlsh> CONSISTENCY QUORUM;
cqlsh> SELECT * FROM cycling_alt.tester where id = 0;
```

Similarly, you can trace the impact of the ONE and ALL consistency levels.

Performance with the ALL consistency setting is worse than that with the QUORUM setting when you're dealing with large data sets or when one node is slower than the rest of the nodes in the cluster.

Ensuring Atomicity with Batch Operations

At times, you may want to process a set of operations (inserts/updates/deletes) as a single all-or-nothing operation. Performing a set of operations so that either all of them succeed, or no operation succeeds, is called an *atomic operation*.

Cassandra allows you to perform *batch operations* in an atomic fashion. For a single partition, Cassandra performs the batch operations without you having to do anything. For batch operations spanning multiple partitions, Cassandra uses a *batchlog*, which means you must configure some additional things.

You can combine multiple INSERT, DELETE, and UDPATE statements into a single operation. Batch operations save on client-server communications, as well as the messaging between the coordinator and the replica nodes.

A batch operation is atomic since if any statement in the batch succeeds, all the statements will succeed, and if one of them fails, all of them will fail. Other transactions in the database can read and write the data that was affected by a partially executed batch operation.

Configuring Batch Operations

You can have Cassandra process batch operations as an all-or-nothing deal. When you use batch operations, you can configure the number of operations in the batch with the `max_mutation_size_in_kb` property. This property determines the maximum size of a single batch mutation (batch operation).

By default, the `max_mutation_size_in_kb` property is set to half of the value you've set for the `commitlog_segment_size_in_mb` property, which sets the size of an individual commit log file segment. The default value of the `commitlog_segment_size_in_mb` property is 32MB. If you decide to set a custom value for the `max_mutation_size_in_kb` property, you must ensure that the value of the `commitlog_segment_size_in_mb` property is at least twice the value of `max_mutation_size_in_kb/1024`.

Note   If the size of the batched operation is greater than the value you've configured for the `max_mutation_size_in_kb` property, the answer isn't to just keep raising the commit log segment size. The problem may lie in an inefficient query access to the data or a wrong data model.

Batch Operations in Single and Multiple Partitions

DataStax points out that multiple partition batch operations suffer from performance issues. You must resort to batch operations for multiple operations only if you have a dire need to ensure atomicity. In a multiple partition batch operation, the coordinator node can turn out to be a bottleneck during a batch operation. The higher the number of partitions in a batch operation, the higher the latency due to batching.

Batch operations that involve writing to multiple partitions also require Cassandra to visit all of those nodes, thus increasing latency of the write operation. As the number of partitions grows, so will the latency of the writes.

The following is an example that shows a batch operation that performs three INSERT operations in one atomic operation. You want all three INSERT statements to work, or none. You start the batch operation with the BEGIN BATCH clause and end it with the APPLY BATCH clause.

```
cqlsh:cycling> begin batch
          ... INSERT INTO cycling.cyclist_expenses (cyclist_name, balance)
              VALUES ('Vera ADRIAN', 0) IF NOT EXISTS;
          ... INSERT INTO cycling.cyclist_expenses (cyclist_name,
              expense_id, amount, description, paid) VALUES ('Vera
              ADRIAN', 1, 7.95, 'Breakfast', false);
          ... iNSERT INTO cycling.cyclist_expenses (cyclist_name, balance)
              VALUES ('Vera ADRIAN',7.95);
          ... apply batch;
 [applied]
-----------
     True
cqlsh:cycling>
```

If a batch operation is successful, you'll see an acknowledgment saying True. If it fails, you'll see an acknowledgment that says the following instead:

```
[applied]
-----------
     False
```

Since all the INSERT statements write to the same partition, this batched write operation is efficient.

When Batch Operations Are Good

Consider batching only if you must ensure atomicity for a set of operations that perform insert/update/deletes. Write operations that use only a single partition are fine performance-wise.

Even in cases with multiple partitions, if the operations involve only small inserts or updates, you'll be fine with batching those operations to ensure consistency.

Lightweight Transactions

Sometimes you may want to read and write data in a sequential order, such as when you are handling banking transactions, where one should carefully handle the credits and debits. Cassandra offers *lightweight transactions* to manage concurrent operations in a transaction. It uses the Paxos consensus protocol to implement lightweight transactions. The Paxos protocol is an algorithm that allows the nodes in a cluster to agree on proposals without requiring a master node to coordinate the transactions. Paxos is an alternative to the traditional two-phase commit used to coordinate distributed transactions.

Relational databases use several types of isolation, including the serializable level of isolation. Cassandra implements the Paxos protocol with linearizable consistency. *Linearizable consistency* yields results similar to those offered by the traditional serializable level offered by relational databases.

Cassandra uses a transaction operation called *compare and set* (CAS) to implement linearizable consistency. Cassandra compares replica data and sets any out-of-date data to the most consistent value in the database.

Performing Lightweight Transactions

An appropriate use case for using a lightweight transaction is when you must insert unique data such as a user's identification number.

Cassandra enables you to issue INSERT and UPDATE statement using the IF clause to support lightweight transactions. For example, you can insert a new cyclist with an ID number, as shown here:

```
cqlsh> INSERT INTO cycling.cyclist_name (id, lastname, firstname)
       VALUES (4647f6d3-7bd2-4085-8d6c-1229351b5498, 'KNETEMANN',
       'Roxxane')
       IF NOT EXISTS;
```

The following statement shows how to perform a lightweight CAS operation against data that exists in the database by adding the IF clause at the end of the operation:

```
cqlsh> UPDATE cycling.cyclist_name
       SET firstname = 'Roxane'
       WHERE id = 4647f6d3-7bd2-4085-8d6c-1229351b5498
       IF firstname = 'Roxxane';
```

How Lightweight Transactions Work

Cassandra intermingles the Paxos protocol and the normal read and write operations. A lightweight transaction doesn't block normal read and operations but it'll block other lightweight transactions.

Since mixing lightweight transactions and normal read and write operations can result in an error, you must use lightweight transactions for both read and write operations. For example, the following set of operations will fail:

```
DELETE ...
INSERT ... IF NOT EXISTS
SELECT ...
```

On the other hand, the following set of operations will work fine:

```
DELETE ... IF EXISTS
INSERT ... IF NOT EXISTS
SELECT ...
```

Caution When Using Lightweight Transactions

Since implementing the Paxos protocol involves a series of actions that occur between proposers and acceptors, there will be multiple round trips between the node that proposes a lightweight transaction and other replicas that are part of the transaction. This obviously has an adverse impact on performance, so you should be careful to use lightweight transactions only when consistency among the operations of a transaction is critical.

Summary

The key takeaway regarding Cassandra data modeling is how queries are the heart of everything. You design your database structures such as tables and indexes based on the queries you expect the database to service.

Understanding the principle of eventual consistency and the various read and write levels and their implications helps you make the best choices to satisfy the requirements of your applications.

CHAPTER 5

Cassandra Architecture

Cassandra has several interesting architectural features that set it apart from a relational database. This chapter introduces the Cassandra architecture. You'll also learn about the Cassandra storage engine and how the database stores its data.

The chapter explains the gossip protocol, which the nodes use to communicate among themselves. Read and write repairs are part of how Cassandra provides consistency. You'll learn about failure detection and recovery. You'll learn about hinted handoff, a feature that ensures that a failed node receives all the modifications that occurred while it was down. You'll also learn about read repair and manual anti-entropy repair.

The chapter explains virtual nodes, which are a significant improvement over traditional ways of distributing data across the nodes of a cluster. You'll learn about various partitioning and replication strategies, such as `SimpleStrategy` and `NetworkTopologyStrategy`, and how to change the replication strategies. You'll learn how to choose among the available data partitioning strategies to help Cassandra distribute data across a cluster's nodes.

Cassandras uses the cluster's topology to decide where it should store the data, as well as how best to respond to queries. To this end, it attempts to store replicas in multiple datacenters to ensure availability. It also sends queries to the local datacenter to minimize latency.

Snitches determine the datacenter and racks that Cassandra writes to and reads from, and are critical for read activity. You'll learn about the several types of snitches offered by Cassandra, such as the `GossipingPropertySnitch` and the `Ec2Snitch`.

Basic Cluster Terminology

Cassandra uses a peer-to-peer system where data is distributed over multiple nodes in a cluster. Nodes in a cluster periodically exchange information among themselves using a *gossip* protocol. One of the cluster's nodes acts as a coordinator or proxy for every client operation and determines which nodes should process the client requests.

© Sam R. Alapati 2018
S. R. Alapati, *Expert Apache Cassandra Administration*, https://doi.org/10.1007/978-1-4842-3126-5_5

Cassandra partitions and replicates data. It uses commit logs to ensure data durability and writes the data from the commit log structures called memtables and from there to disk in SSTable data files.

Periodically, the database compacts data by discarding obsolete data. To ensure consistency of data, Cassandra employs several types of repair mechanisms.

It's a good idea to refresh your understanding of some of the key terms you'll come across often in this and the rest of the chapters when dealing with the architecture of Cassandra.

- A *datacenter* is a group of related nodes that you configure together within a cluster for replication purposes. You can have virtual or physical datacenters. A datacenter can't span physical locations. The purpose in having multiple datacenters is to prevent other data nodes from adversely impacting transactions in a data node and to lower latency. You can create datacenters to contain specific node types, such as transactional, analytics, search, and graph.

- A *rack* is a logical set of nodes located close to each other, usually in the same physical rack of physical machines. A datacenter is a logical grouping of racks, connected by a network.

Note   Cassandra comes with a default rack named RAC1 and a default single data center named DC1.

- A *cluster* is the largest unit of deployment for a Cassandra database and consists of a set of nodes in one or more locations. In AWS, the locations are multiple availability zones. A cluster can consist of just a single node, a single datacenter, or multiple datacenters. A cluster is sometimes referred to as a *ring* since Cassandra stores data in the cluster by arranging the data in a ring.

- A Cassandra *node* is a part of an individual location (a server) on a Cassandra cluster that stores data. To be specific, it stores data partitioned according to a partitioning algorithm that you specify.

- A *commit log* is a write-ahead transaction log that Cassandra stores on every node of the cluster. Cassandra writes data first to the append-only commit log, which it then flushes to disk in batches or periodically. Commit logs perform a critical role during a data recovery. Once the database flushes its data to disk, the database can archive, delete, or reuse the commit log.

- A *memtable* is an in-memory structure that's a cache of data rows that can be looked up by a key. Each memtable stores data for a specific table and the database flushes the memtable contents to disk when the commit log gets full or after a specific interval (set at the table level).

- An *SSTable* (sorted strings table) is a logical entity that is comprised of several files on disk. An SSTable is immutable and is an append-only structure. Cassandra creates an SSTable when it flushes a memtable to disk. Cassandra sorts the memtables by key and writes them out sequentially to create an SSTable. This is the reason Cassandra's writes are very fast; they involve only the appending of a commit log and a sequential write operation for flushing the data.

- A *keyspace* is a logical container that contains one or more tables and is analogous to a schema in a relational database. You store data inside tables that are part of a specific keyspace. All data objects such as tables must belong to a keyspace.

  - A keyspace defines how Cassandra replicates data on the nodes of a cluster. A keyspace has several storage attributes. The keyspace also determines the RF (replication factor) for data that you store in a keyspace. You define the data replication strategy at the keyspace level, either when you create it or later.

  - It's a good idea to have a separate keyspace per application.

- A CQL table (aka *column family*) is a logical entity that resides in a keyspace. A table consists of ordered columns that the database fetches by row. You must define a primary key when you create a table.

Replica Placement Strategy

Cassandra allows you to replicate data by storing multiple copies of data across the nodes of a cluster. There are two strategies you can adopt to let Cassandra know how it ought to replicate data:

- The *SimpleStrategy* is for prototyping.

- The *NetworkTopologyStrategy* is for production environments.

How Cassandra Stores Data

Since Cassandra is a key-value store, it organizes its data into rows that consist of several columns that represent the values. A row belongs to a single node, and Cassandra replicates that row to the other nodes in the cluster.

Cassandra hashes each row's key to determine in which node in the ring it should store that data. It organizes the cluster itself as a ring of nodes, with each node storing an equal number of hash values.

Let's say you are running a four-node Cassandra cluster. You have a single keyspace and you choose a replication factor of 3. Let's say you create a table in the keyspace and insert data into it. Each time Cassandra inserts data, it hashes the value of that row's key to find out which node is responsible for storing that row.

For simplicity's sake, let's assume the hash values range over 1-100, although in reality the hashes use 128-bit or larger values. Since you have four nodes in the cluster, here's how the nodes divide the hash values among themselves:

- Node 1: Hash values 1-25

- Node 2: Hash values 26-50

- Node 3: Hash values 51-75

- Node 4: Hash values 76-100

Figure 5-1 shows how Cassandra distributes the hash values among the four nodes.

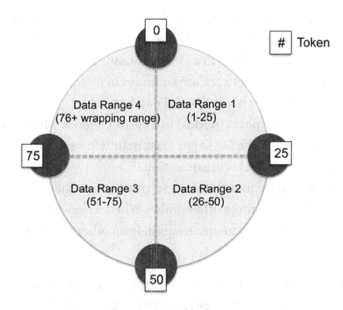

Figure 5-1. *How nodes are responsible for various hash values*

Let's say that the first row you insert gets hashed by Cassandra and the hash has a value of 65. As the diagram shows, the row belongs to Node 3, and Cassandra will also replicate the data to Node1 and Node 2. Why nodes 1 and 2? The reason is that replication is always done on a clockwise basis in a ring.

Let's say that the next row you insert has the value 48. It goes to Node 2, and Cassandra replicates the row to Node 3 and Node 4.

This is the essence of how Cassandra stores data among the nodes in a cluster.

Ideally, all nodes in a cluster have an equal share of the hash values and thus the data set. That's why it's very important to ensure that the primary key you select for a table has a high cardinality (many values). This way, the rows are spread among all the nodes in the cluster instead of being stored in a skewed fashion, with some of the nodes having the responsibility for a higher portion of the data set than others. This results in hot spots on the storage drives, as fewer nodes are doing most of the work during read operations.

Gossip and How It Helps Nodes Communicate

Cassandra uses *gossip*, a protocol in which nodes exchange state information about themselves and other nodes they're aware of in order to discover location and state information about all nodes in a cluster. Gossip protocols that assume a less-than-perfectly-functioning network are common in distributed network systems and are used for data replication in distributed databases.

As with human gossip, the peers (nodes) in a cluster choose which nodes they want to exchange information with. The `Gossiper` class maintains a list of nodes that are alive, thus enabling gossip to help with failure detection.

The gossip process runs frequently (every second) and exchanges state messages of one node with those of up to three other nodes. When a node starts, it registers with the gossiper to receive endpoint state information from other nodes in the cluster, as shown here:

```
INFO  [main] 2017-06-24 12:19:21,997 MessagingService.java:733 - Starting
Messaging Service on ubuntu/192.168.159.130:7000 (ens33)
INFO  [HANDSHAKE-/192.168.159.129] 2017-06-24 12:19:22,199
OutboundTcpConnection.java:510 - Handshaking version with /192.168.159.129
INFO  [GossipStage:1] 2017-06-24 12:19:22,326 Gossiper.java:1056 - Node
/192.168.159.129 is now part of the cluster
INFO  [RequestResponseStage-1] 2017-06-24 12:19:22,380 Gossiper.java:1020 -
InetAddress /192.168.159.129 is now UP
INFO  [main] 2017-06-24 12:19:23,202 StorageService.java:705 - Loading
persisted ring state
INFO  [main] 2017-06-24 12:19:23,218 StorageService.java:818 - Starting up
server gossip
INFO  [GossipStage:1] 2017-06-24 12:19:24,596 Gossiper.java:1056 - Node
/192.168.159.129 is now part of the cluster
INFO  [RequestResponseStage-1] 2017-06-24 12:19:24,601 Gossiper.java:1020 -
InetAddress /192.168.159.129 is now UP
INFO  [main] 2017-06-24 12:19:31,610 CassandraDaemon.java:725 - No gossip
backlog; proceeding
```

And when a node can't get the gossip information, it lets you know about that too, by throwing an exception:

```
INFO  [main] 2017-06-24 12:06:00,302 MessagingService.java:733 - Starting
Messaging Service on
ERROR [main] 2017-06-24 12:06:31,476 CassandraDaemon.java:752 - Exception
encountered during startup
java.lang.RuntimeException: Unable to gossip with any seeds
WARN  [StorageServiceShutdownHook] 2017-06-24 12:06:31,564 Gossiper.
java:1514 - No local state, state is in silent shutdown, or node hasn't
joined, not announcing shutdown
```

The gossip process can be both direct (other nodes gossip to a node) and indirect (nodes get information from other nodes, which might get it from still other nodes). Since each node not only exchanges information about its own state but also about the state of the other nodes they communicated with, all nodes will be aware of the state of all the other nodes in the cluster very fast.

When nodes exchange gossip among themselves, the latest state of a node overwrites older state information.

When the gossiper finds that an endpoint isn't responding, it marks that endpoint as dead on its list and logs the information.

Cassandra uses a sophisticated algorithm called *Phi Accrual Failure Detection* for detecting node failures. Instead of using simple heartbeats from the nodes to indicate that they are alive, the algorithm uses a *suspicion level* to determine node availability.

Whereas a heartbeat offers a simple yes/no criterion to determine if a node is alive, the failure detection uses a continuum of data to make that determination. For example, if a node can't accept a connection occasionally due to blips in the network, that node isn't immediately put on the list of dead nodes. Thus, the algorithm matches real-life network conditions and realistically portrays the health of the nodes. Using this algorithm, the database detects node failures within 10 seconds after they occur.

Configuring Gossip Settings

You really don't need to do anything to set up gossip because all the necessary properties are set when you start up the Cassandra nodes in a cluster. When you start up a node, you configure the following properties, all of which help the node know which nodes it should contact for gossip about other nodes. These properties also set the ports and other relevant IP addresses or hostnames to enable Cassandra to connect the nodes to one another.

- `cluster_name`: Name of the cluster the node is a part of.

- `listen_address`: The IP address/hostname that Cassandra must bind to so it can connect this node to the rest of the nodes in the cluster. Alternatively, you can specify the `listen_interface` property instead of `listen_address`.

Tip   Specify multiple seed nodes per datacenter so gossip doesn't have to communicate with nodes from a different datacenter when bootstrapping nodes.

- `seed_provider`: This property enables you to specify a list of hosts (IP addresses) that gossip relies on to learn the ring's topology. You don't need to specify all nodes as seed nodes since it reduces gossip performance.

- DataStax recommends that you use a small number of seeds, such as three nodes per datacenter. You should specify the same list of nodes for all nodes in a cluster. If your cluster has multiple datacenters, include at least one node from each datacenter as seed providers.

- `storage_port`: This is the inter-node communication port and it must be identical across all nodes in the cluster.

Seed Nodes and Gossip

In Chapter 3, you learned about the seed nodes that you specify in the `cassandra.yaml` file. Seed nodes aren't really a big deal, in the sense that their failure doesn't constitute a single point of failure for the cluster. The only role they play is in bootstrapping the gossip process for a new node that's joining the cluster.

You must ensure that you specify an identical set of seed nodes for all of a cluster's nodes. This point is quite important when you first start up a node.

It's not advisable to make every node a seed node; this will only adversely impact gossip performance. Ideally, your seed list should be small, with about three nodes per datacenter. If you have multiple datacenters, include at least one node from each datacenter in the seed list.

If you include two or three nodes per datacenter gossip, as against specifying a single seed node, it protects you against the failure of the single node. Otherwise, when the single seed node fails, gossip will need to communicate with a different datacenter during the bootstrapping of the nodes.

Failure Detection and Recovery

Cassandra determines the status of a node (up/down) from the gossip state and does its best to avoid sending client requests to nodes that are unreachable.

Cassandra uses an *accrual detection mechanism* to determine the threshold for marking a node as failed. This threshold considers network performance, workload, and historical data to help calculate the critical threshold for node unavailability.

All nodes use a failure detection mechanism to detect the failure of the rest of the nodes in the cluster. Nodes use the latency of the gossip messages by using a sliding window on the arrival times of gossip messages from other nodes.

You can configure the failure detection mechanism by adjusting the `phi_convict_threshold` property in the `cassandra.yaml` file. The default value for this property is 8. If you set this property's value at a high number, it's possible the cluster will miss transient failures. A lower value, on the other hand, increases the odds of marking an unresponsive node as a down node.

Amazon EC2 experiences frequent network congestion issues. In this and other environments where you may experience unusable networks, you can raise the value of the `phi_convict_threshold` property to 10 or 12 to avoid misleading failure detection. In general, according to DataStax, a range of values between 5 and 12 is good enough in most cases.

Intermittent node failures such as those caused by network outages are transient and don't result in the removal of the node from the ring. The rest of the nodes will periodically try to contact the unreachable node.

When an unreachable node comes back online, Cassandra uses repair mechanisms such as hinted handoffs and manual repair to ensure that the node catches up with the missed writes for the replica data it maintains.

Repairing a Node

Cassandra is a distributed database and over time the data in one of the replicas can diverge from that of the other replicas. *Node repair* is a process that corrects the inconsistencies among the replicas so eventually all nodes will have identical data. As an administrator, node repair is one of your key responsibilities and is something that you'll perform on a regular basis.

Cassandra offers three types of repairs, some of which involve your manual intervention:

- Hinted handoff

- Read repair

- Anti-entropy repair

The following sections briefly explain the various types of repair.

Hinted Handoff

Hinted handoff is a process where a node that's acting as the coordinator for a write that it was unable to perform on a different node saves the data as a set of *hints*. When the unreachable node becomes available later, the coordinator node hands off the stored hints to help the other node catch up with the missed writes.

A hint contains data about the write request that was sent to the node that was unavailable at that time. A hint is like a wrapper around the change operation that tells the database that it needs to replay the data to the node that was unavailable at the time of the write.

The database creates a separate hint for each partition to which it must write data. Once the coordinator node finds from gossip that the failed node is back up again, it hands off the hint to the revived node to write the missed data. Hints thus help the time it takes for a failed node to become consistent when it comes back up.

The idea behind hinted tradeoff is to enable a cluster to operate normally by enabling it to perform its normal amount of writes even when one or more nodes are unavailable. Thus, this feature directly contributes to Cassandra's vaunted high availability. Hints allow Cassandra to offer full write availability when consistency isn't a requirement. They also enhance response consistency after temporary outages such as a network failure.

Tip   You can disable hinted handoff to avoid a situation where a node may be down for long. This will keep the database from having to send a flood of hints to the failed node when it comes back online.

A hint doesn't count towards a write when satisfying the consistency level. However, the consistency level ANY accepts a hint as a something that satisfies the requirements for a read operation. Even if the database were to write just a hint, the write would be successful.

What a Hint Is

Cassandra stresses all hints on the local hints directory on each node. Besides the data itself, each hint consists of a few identifying attributes for the hint. Here's what a hint consists of:

- The ID of the down or unavailable node

- A hint ID, which is a time UUID for the data

- A message ID with the Cassandra version

- The data in the form of a blob

The database flushes all hints to disk every 10 seconds to keep the hints fresh. Once the down node comes back up, Cassandra writes all data to the node and deletes the hint file after that.

When Cassandra attempts to write to a replica node that it knows is down or is not responding to a write request, the coordinator node stores a hint on the node where it's running, in the `system.hints` table. When the gossip mechanism lets the coordinator node know that the failed node is back, the coordinator node sends the data rows corresponding to each of the hints to the revived node.

For brief outages, the coordinator node checks every ten minutes for any writes that may have been missed due to an intermittent failure and that weren't caught by failure detector via gossip.

How Hinted Handoff Works

For hinted handoff to work, first you must enable it. Since the default value for the `hinted_handoff_enabled` property in the `cassandra.yaml` file is `true`, you don't need to do anything, unless you disabled this feature earlier for some reason.

You can also enable hinted handoff on a per datacenter basis by passing the list of datacenters, as shown here:

```
hinted_handoff_enabled: DC1, DC3
```

You can specify a list of datacenters that won't perform hinted handoff; that is, disable the feature in some of the datacenters in a cluster by passing the list of the blacked-out datacenters as values to the `hinted_handoff_disabled_datacenters` property, as shown here:

```
hinted_handoff_disabled_datacenters: - DC1 - DC2
```

Managing a Hinted Handoff

By default, the database stores the hints in a directory named `hints` in the following location:

```
$CASSANDRA_HOME/data/hints
```

You can set a different location for the `hints` directory by setting the `hints_directory` property.

You can configure the maximum size of a single hints file (in megabytes) by setting the `max_hints_file_size_in_mb` property (default is 128MB). You can have the database compress the hints file by setting the `hints_compression` parameter, which defaults to LZ4Compressor. Besides LZ, the database also supports the Snappy and Deflate compression algorithms as compressors for a hints file.

Checking the Status of a Hinted Handoff

You can find out the status of hinted handoff by issuing the `nodetool statushandoff` command:

```
$ nodetool <options> statushandoff
```

The options included the hostname, port number, and password. Here's an example:

```
$ nodetool statushandoff
Hinted handoff is running
$
```

Removing All Hints

You can remove all hints on the local node, or for one or more endpoints, by using the `nodetool` `truncatehints` command:

```
$ nodetool <options> truncatehints -- <endpoints> ...
```

The `endpoints` options enable you to specify one or more IP addresses or host names where you want to delete the hints.

Disabling and Enabling Hints for a Datacenter

You can enable hints for a datacenter with the `nodetool enablehintsfordc` command, which has the following syntax:

```
$ nodetool [options] enablehintsfordc [--] <datacenter>
```

The `datacenter` option is how you specify the datacenter for which you want to enable hints. Here's an example of how you invoke this command:

```
$ nodetool -u username -pw password enablehintsfordc DC1
```

Earlier, I explained how you can configure the `hinted_handoff_disabled_datacenters` property to blacklist one or more datacenters when you start up a cluster. You can run the `nodetool enablehintsfordc` command to enable hints for the datacenters you've backlisted at startup.

As you might guess, you can disable hints for a datacenter by running the `nodetool disablehintsfordc` command, as shown in the following example:

```
$ nodetool -u username -pw password disablehintsfordc DC1
```

You usually disable hints for a specific datacenter when the datacenter is down but the rest of the cluster is working fine. You can also do this when failing a datacenter over.

Tuning Hinted Handoff

You can control the rate at which the surviving nodes send hints to a recovered node. You do this by setting the `sethintedhandoffthrottlekb` property (in KB/second). The property doesn't set the rate of transmission of the hints from the good nodes to the recovering node. Rather, it sets the maximum sleep interval per delivery thread after it delivers a hint.

You set the `sethintedhandoffthrottlekb` property in the following way:

```
$ nodetool sethintedhandoffthrottlekb 4096
```

This example shows how to set the hinted handoff throttle to 4096 KB/second per delivery thread.

The more nodes you have in your cluster, the smaller the sleep interval between hints because all the delivery threads from the various nodes share the maximum interval. The more nodes there are, the more the simultaneous delivery of hints, hence the lower the real sleep interval between the transmissions of the hints.

You can speed up hint handoffs in a multiple datacenter deployment by configuring the `max_hints_delivery_threads` property, whose default value is 2. By raising this value, you enable the database to use a higher number of threads to deliver hints during a cross-datacenter handoff of hints.

Stopping the Writing of New Hints

When a node is down for a long time, the coordinator node won't write anymore new hints. You can configure this period with the `max_hint_window_in_ms` attribute in the `cassandra.yaml` file. By default, this period is set to 3 hours.

The reason Cassandra imposes a time interval for the period for which the database generates hints is to ensure that a large volume of stored hints won't cause a resource crunch when the database tries to write all the data to the node that was brought up after a prolonged period of time.

The `write_request_timeout_in_ms` property specifies the time (in milliseconds) that the coordinator will wait for write operations to complete, before it times out the operation. The default is 2000 milliseconds.

Read Repair

Cassandra performs a *read repair* when it repairs nodes that are queried during a read operation. The extent of the read repair depends on the replication factor you've configured for a database. If you've configured a CONSISTENCY level of ONE or ANY, there's only a single replica node that's queried by the coordinator node. However, if you've set the CONSISTENCY level to greater than ONE, there are multiple replica nodes that the coordinator will get in touch with for every read request.

During a read, the coordinator sends out requests to one of the nodes that has a replica (replica node). Simultaneously, it also sends digest requests to all the other replica nodes where you've configured a CONSISTENCY level greater than ONE.

The coordinator node returns the requested data to the client if all of the nodes return the same (consistent) data. However, if some of the replica nodes don't have the same data, the coordinator writes the latest version of the data to the replica nodes with the outdated replicas.

Random Read Repairs

Cassandra can choose to randomly perform a read repair on its own of all the replicas of a table. The replication factor has no bearing on this type of random read repair.

The read_repair_chance property that you set for a table determines how frequently the database performs random read repairs. This property sets the probability that a successful read operation will trigger a read repair. The default value for this property is zero, and you must set it to a value between zero and one.

Whereas the read_repair_chance property determines the probability of a read repair across the cluster, a related table level property, dclocal_read_repair_chance, determines the probability of a read repair of replicas in the same datacenter as the coordinator.

The recommended value for the read_repair_chance property is a value of 0.2 for all compaction strategies except the deprecated DataTieredCompcationStrategy, which you set to zero. However, your mileage may vary so proceed with caution when setting a value for this property.

Rapid Read Protection

Cassandra's *rapid read protection* feature allows it to tolerate a node failure without losing even a single request. If a node is lost, the dead node can't service any client requests until failure detection occurs. However, if you enable rapid read protection, there's only a brief dip in the read traffic.

Why Rapid Read Protection Helps

Cassandra employs snitches to send requests to the replicas with the least load (replicas that are likely to respond the fastest to a read request). Cassandra sends normal read requests to just enough nodes to satisfy the consistency level.

Cassandra performs a different number of read requests in each case, with the number of requests dependent on the consistency level you configure. For example, the following are the number of read requests Cassandra performs for various consistency levels:

- Consistency level ONE: One read request

- Consistency level QUORUM: Two read requests

Figure 5-2 shows how the coordinator sends read requests to the nodes.

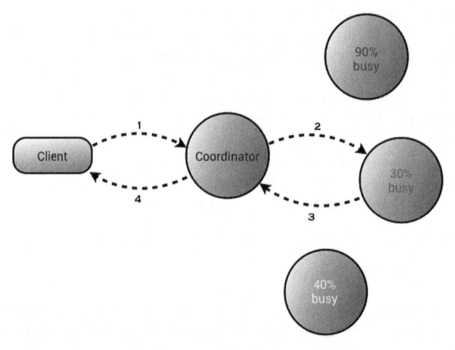

Figure 5-2. *How the coordinator sends read requests to the "best" node in a cluster*

Note   The speculative_retry property enables you to configure a trigger for firing extra read requests when the requests aren't being satisfied in the expected time.

Rapid read protection enables the coordinator to send redundant requests to other replicas when the target replica is slower than expected, by monitoring the outstanding read requests. That is, the database sends out extra read requests to other replica nodes even after it meets the consistency level requirements. Rapid read protection helps during node failures, or other events that slow a node's throughput, or a throughput that's adversely affected by a cluster-wide, full, un-throttled compaction.

Configuring Rapid Read Protection

You configure rapid read protection by setting the `speculative_retry` property at the table level, either when you create it or later. The `speculative_retry` property overrides the normal read timeout (when you set the `read_repair_chance` property to a value other than 1.0) and sends out another read request.

The `speculative_retry` property can take the following values:

- `ALWAYS`: The coordinator node always sends extra read requests to all replicas after every read of this table.

- `NONE`: The coordinator node never sends extra read requests after a read of this table.

- `Xpercentile`: The coordinator node sends extra read requests if the table's latency is higher than normal. That is, extra reads are triggered after a percent of the normal read latency of a table elapses. For example, the coordinator sends out extra read requests after waiting for 48ms (80 percent of 60ms) if a table's latency is 60 seconds on average.

- `Nms`: The coordinator node sends out extra read requests if it doesn't receive any results from the target node in N milliseconds.

The default value of the `speculative_retry` property is `99percentile`. You can set the `speculative_retry` property when creating a table as shown here where you use `3ms` as the criterion for triggering additional read requests:

```
speculative_retry = '3ms'
```

You can also configure rapid read protection after creating a table:

```
cqlsh> ALTER TABLE users WITH speculative_retry = '10ms';
cqlsh> ALTER TABLE users WITH speculative_retry = '90percentile';
```

Since a consistency level of ALL requires responses from all replicas, you can't use rapid read protection.

In general, the more nodes in your cluster, the smaller the impact of rapid read protection on the nodes. This is a good benefit of spreading replication throughout the cluster with the use of virtual nodes.

Anti-Entropy Repair

Data can become inconsistent over time due to frequent data deletions and node crashes. In these situations, you need to maintain the nodes by performing *anti-entropy repair*. You perform this repair manually with the `nodetool repair` command. Anti-entropy repair is a regular maintenance task for administrators.

How Anti-Entropy Repair Works

During anti-entropy repair, Cassandra compares all replicas and updates those replicas to the latest version of the data. Cassandra uses Merkle trees (binary hash trees) so it can check data independently without the coordinator node having to download the entire data set.

When the initiating node detects differences in the Merkel trees from participating peer nodes, it exchanges data for the configuring ranges and Cassandra writes the new data to the SSTables.

You can run the `nodetool repair` command on a single node or on all the nodes of a cluster. The node that initiates the anti-entropy repair serves as the coordinator node for the repair operation.

Although you can have Cassandra perform a full repair of a node's data by comparing all SSTables of a node and make the repairs, by default the database performs incremental repair. Incremental repair relies on metadata that shows the SSTable's rows as repaired or unrepaired. It persists already repaired data and builds Merkle trees only for those SSTables that it hasn't repaired yet.

Performing a Manual (Anti-Entropy) Repair

You must perform an anti_entropy node repair regularly, especially when there are frequent deletes in the database. The purpose of the anti-entropy repair command is twofold:

- Ensure data consistency on all of the cluster's nodes.

- Repair data inconsistency on a node that was down.

Use the nodetool repair command to repair one or more tables. When you run a full repair, Cassandra marks the SSTables as repaired. Run the nodetool repair command to perform routine maintenance and not to repair a node that was down.

Caution   You shouldn't run the nodetool repair command after making any topography changes in your cluster.

The nodetool repair command has several parameters, as you can see here:

```
$ nodetool [(-h <host> | --host <host>)] [(-p <port> | --port <port>)]
    [(-pw password | --password password)]
    [(-pwf passwordFilePath | --password-file passwordFilePath)]
    [(-u username | --username username)] repair
    [(-dc specific_dc | --in-dc specific_dc)...] [(-pl | --pull)]
    [(-dcpar | --dc-parallel)] [(-et end_token | --end-token end_token)]
    [(-full | --full)]
    [(-hosts specific_host | --in-hosts specific_host)...]
    [(-j job_threads | --job-threads job_threads)]
    [(-local | --in-local-dc)] [(-pr | --partitioner-range)]
    [(-seq | --sequential)]
    [(-st start_token | --start-token start_token)] [(-tr | --trace)]
    [--] [keyspace tables...]
```

In the following sections, I show how to configure the most important parameters when running the nodetool repair command.

Full vs. Incremental Repair

You can configure `nodetool repair` with options that let you perform the repair in a full or an incremental fashion.

The default mode for repair is incremental and it performs better than a full repair, since the database needs to worry about calculating the Merkle tree only for those SSTables that it didn't repair earlier. That is, the database skips the SSTables marked as repaired. By performing frequent incremental repairs you can keep the repair process short and efficient.

You can perform a full repair by specifying the `-full` option. The `-hosts` option lets you specify the good nodes to use for repairing any bad nodes, which you specify with the `-h` option.

A full repair prevents anti-compaction, which is the process of splitting an SSTable into two tables, with one table containing the repaired data and the other the original data.

Sequential vs. Parallel Repair

If you want to complete a repair quickly and you've sufficient resources for the work, you can choose to perform a parallel repair, which means that the database repairs all data centers in parallel.

Specify the `-dcpar` option if you want to do a parallel repair; you may not actually have to do this because parallel repair is the default. The following command performs a parallel repair:

```
$ nodetool repair
```

The alternative to a parallel repair is sequential repair. The following `nodetool repair` command performs a sequential repair of all keyspaces on this node:

```
$ nodetool repair -seq
```

Partitioner Range Repair

Since Cassandra replicates data, you may end up making the database repair the same data range more than once by performing the repair one node at a time. You can conserve resources by having the `nodetool repair` command repair a specific range of data just once.

You can perform a partitioner range repair by specifying the -pr option. While this option is a good one, you can't combine it with an incremental repair, which is the default mode. The following example shows how to perform a partitioner range repair of a bad partition on this node by using the good partitions on either node 10.2.2.30 or node 10.2.2.31:

```
$ nodetool repair -pr hosts 10.2.2.30 10.2.2.31
```

Tip   DataStax recommends that you use the parallel and partitioner range options during a repair wherever it's possible to do so.

Restricting the Repair to the Local Datacenter (vs. a Cluster-Wide Repair)

By default, the nodetool repair command runs the repair process cluster wide, that is, on every node that contains replicas regardless of the datacenters. If you have three datacenters and a replication factor of 3, you're asking the repair process to build nine Merkel tables, causing your network and other resource usage to spike.

You can limit repairs to specific datacenters or just the local node where you run the repair command. Specify the -dc or –in–dc options to limit repairs to specific datacenters. The following example shows how to restrict the node repair just to the local datacenter *DC1* by specifying the -dc option:

```
$ nodetool repair -dc DC1
```

Endpoint Range vs. Subrange Repair

By default, Cassandra repairs all partition ranges on a node or endpoint range. You can alternatively run a subrange repair by specifying the -st (or -start-token) and the -et (or -end-token) options.

A subrange repair enables you to target partition ranges precisely to fix errors but it requires generated token ranges.

When and How Often to Perform an Anti-Entropy Repair

There are several situations that call for a node repair, as I summarize here.

- Run it as a regular maintenance operation, especially when there are frequent deletes in the database.

- Repair the nodes where you have data that's rarely read, since they don't undergo read repair.

- Run repairs on a node that you've recovered after a failure.

- You must also run repair to recover missing data or when an SSTable is corrupted.

Ideally, you should run incremental repairs every day and a full repair less frequently, like every month, unless you believe you need to do it more often. A full repair is useful even when you don't have any deletions of data since it helps maintain data integrity.

The table property `gc_grace_seconds` has a lot to with how frequently you repair a node. The `gc_grace_seconds` table option specifies the length of time after the database marks data with a deletion marker (called a *tombstone*) before making it eligible for deletion. By default, the interval is 24 hours (864,000 seconds), which allows the database time to maximize consistency before it deletes the data for good.

The reason Cassandra delays the garbage collection (through configuring the value for garbage collection grace seconds by setting the `gc_grace_seconds` property) is to let an unavailable node have sufficient time to recover. If the node doesn't come up within this period, the database will consider it a failed node.

Since the default value of the `gc_grace_seconds` property is 24 hours, it's a good idea to run the repair daily so as to properly handle all deleted data. You can set a longer interval for the `gc_grace_seconds` property for data that's rarely deleted or modified. In this case, you can perform the repair less frequently for these tables.

Migrating to Incremental Repair

Although incremental repair is much more efficient than a full repair, the process could take a long time the very first time because Cassandra needs to recompact all the SSTables. DataStax recommends that you migrate your cluster one node a time to incremental repair. The following steps show how to do the migration.

1. Disable autocompaction.

    ```
    $ nodetool disableautocompaction
    ```

2. List the node's SSTables in a text file, such as `SSTable-names.txt`. You'll find the SSTables under the `/var/lib/cassandra/data` directory. In the `data` directory, there's a directory for each keyspace and it contains a list of files for each SSTable in that keyspace. You need to list all the files with the SSTable data. These files have the following format:

    ```
    <version_code>-<generation>-<format>-Data.db
    ```

    The following command helps you list all the `Data.db` files in a keyspace:

    ```
    $ find '/homeuser/Datastax-ddc-3.2.0/data/keyspace1'
    -iname "*Data.db*"
    ```

    The SSTable file list (`.db` files) you generate will look like the following:

    ```
     /data/cycling/cyclist_by_country-82246fc065ff11e5a4c58b49
    6c707234/ma-1-big-Data.db
    /data/cycling/cyclist_by_birthday-8248246065ff11e5a4c58b49
    6c707234/ma-1-big-Data.db
    /data/cycling/cyclist_by_birthday-8248246065ff11e5a4c58b49
    6c707234/ma-2-big-Data.db
    ```

    Gathering the SSTables list in a text file helps you perform a batch process to mark all SSTables in Step 5 as repaired.

3. Run a full repair on just this node.

    ```
    $ nodetool repair
    ```

    By default, the repair is full and sequential (not parallel).

4. Shut down the node.

5. Set the `repairedAt` flag on all SSTables to `-is-repaired` using the
    `sstablerepairedset` command. Before you can run incremental
    repair, it is mandatory that you set the repaired status of the SSTable.

    ```
    $ sudo sstablerepairedset –really-set –is-repaired -f SSTble-names.txt
    ```

    The `sstablerepairedset` tool sets the `repairedAt` status of one or
    more SSTables by marking the tables as *repaired* or *unrepaired*. In
    this case, you used a file with a list of SSTable files, but you could
    also specify an individual SSTable to mark the repair status of a
    single table.

6. Once the repair is complete, restart the node.

Once you migrate all the nodes in your cluster, you can perform incremental repairs
(with the `nodetool repair -inc` command) on each of the nodes.

Data Distribution and Replication

A *replica* is a copy of a table row. When Cassandra first writes data to a table, it refers to it
as a replica as well.

Data distribution and replication are key topics that affect Cassandra performance.
To understand how Cassandra distributes data and replicates copies of that data across
the cluster, you need to understand the following topics:

- Virtual nodes

- Replication strategies

- Partitioners and partitioning strategies

- Snitches

In the following sections, I explain each of these crucial elements of Cassandra data
distribution and replication.

Virtual Nodes and Data Ownership

You can use either a single-token architecture or virtual nodes to determine how Cassandra distributes data across the nodes in a cluster.

The database represents the data in a cluster as a ring, with each node responsible for a chunk of the ring of data. Cassandra assigns each node one or more ranges of data, described by a token. A token is a 64-bit integer ID that helps identify the partitions in the ring. The tokens can range from -2 to the power of 63 to 2 to the power of 63 -1. The tokens determine the position of the nodes in the ring.

Each node in the ring owns a range of values that is greater than the token of the preceding node and less than or equal to its token. This way the token ranges are spread across the nodes that live in different racks.

Virtual nodes, simply called *vnodes*, help you distribute data at a more granular level than by calculating tokens. Vnodes automatically calculate and assign tokens to each of the nodes in a cluster. This eliminates the need for you to determine the partition ranges by calculating and assigning tokens.

In a non-vnode environment, there's just one token per node, and most of the time a node owns just one contiguous range in the ringspace, and there's often one range per node. Vnodes allow multiple tokens or ranges that are randomly selected and are non-contiguous, thus giving you many small token ranges that belong to each node.

Each of the virtual nodes owns a portion of the token space. Cassandra randomly distributes the virtual nodes among the physical nodes; a physical node doesn't own a contiguous range of virtual nodes. Cassandra recommends 256 virtual nodes in a cluster, so if you have 16 nodes in your cluster, each of the physical nodes will be assigned 16 virtual nodes on a random basis, with each virtual node responsible for a different token range.

Since you always replicate data in a Cassandra database, it means that the database stores each of the virtual nodes on multiple nodes.

Vnodes make it easy to rebalance data across the cluster when you add or remove nodes. New nodes automatically assume responsibility for an even amount of data from the rest of the nodes. Similarly, when a node goes out of the cluster, the rest of the nodes divide its data amongst themselves. A dead node is rebuilt quickly since all nodes in the cluster help distribute its data across themselves.

Vnodes help you employ heterogeneous hardware because they allow you to vary the proportion of vnodes assigned to each node based on the physical capacity of the node.

You can use different token architectures across a cluster, with different datacenters employing different architectures. That is, some data centers may use vnodes, and other data centers, no vnodes. However, all nodes in a single datacenter must either be either vnode-enabled or use a single-token architecture.

How Data Is Distributed Across a Cluster

When you create a datacenter, the database evenly distributes the workload among the nodes. Over time, as you add and/or remove nodes, the topology of the datacenter changes and the workload could become unbalanced.

A token assigns a range of data to a specific node that's a part of a Cassandra cluster. The token or tokens of a node determine the node's range of data. Each of the nodes is responsible for the ring's region between itself and its predecessor.

Let's say the range of tokens is 0 to 100, and you have four nodes in the cluster. The tokens for the nodes will be 0, 25, 50, and 75 in this case, assuring that each node is responsible for an equal range of the data.

In earlier releases (prior to Cassandra 1.2), you had to calculate and assign a single token to each node. Figure 5-3 shows this architecture.

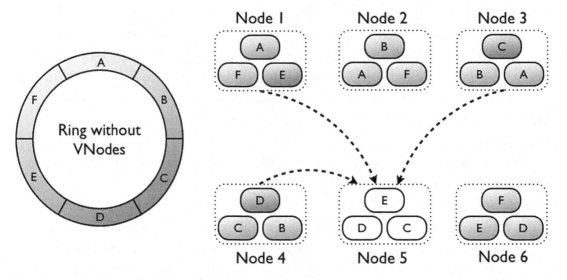

Figure 5-3. *A single-token architecture*

As Figure 5-3 shows, each of the six nodes, Node1–Node6, is assigned a single token, A, B, C, D, E, and F. The single token represents a location within the ring. Each node will have the data that is determined by mapping the partition key to a token value within the range of data assigned to each node.

Tip   Nothing prevents you from setting a replication factor greater than the number of nodes in a cluster. Normally it isn't a smart idea to do this because it really doesn't offer enhanced reliability. However, in some circumstances, you can do this in anticipation of adding more nodes later on.

Assuming a replication factor greater than one, in addition to the range assigned to it, each node will also store copies of rows from the other nodes. For example, assuming a replication factor of four, Figure 5-4 shows how each node stores data from ranges other than the single range assigned to it. In addition, each of the nodes occupies a contiguous partition range in the ring space.

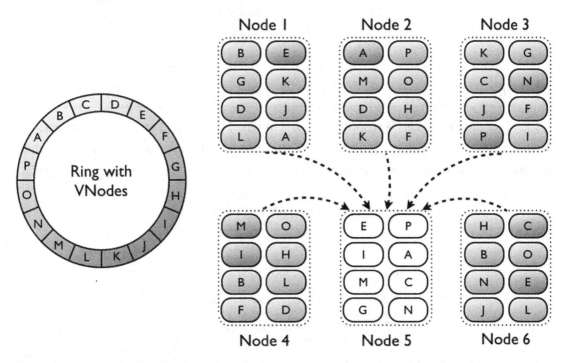

Figure 5-4. *A ring with virtual nodes*

Virtual nodes allow each node to own multiple smaller partition ranges distributed in the cluster instead of a single token. As with single tokens, Vnodes use consistent hashing to distribute data across the cluster without generating and assigning tokens. Figure 5-4 shows a ring with virtual nodes.

As Figure 5-4 shows, virtual nodes are non-contiguous and are randomly assigned to the nodes. Cassandra determines to which partition range it must assign a table row by the hash of the partition key.

Consistent Hashing

Cassandra uses constant hashing to partition data according to the partition key. This enables the cluster to distribute data across the nodes with a minimal reorganization when you add or remove nodes.

Let's say you have a four-node cluster with the following data in a table:

Name	Age	Car	Gender
Jim	36	Camaro	M
Carol	37	BMW	F
Johnny	12	Ferrari	M
Suzy	10	Ford	F

Each of the four nodes is responsible for a range of data based on the hash values. Assuming you're using the default Murmur3 partitioning strategy, Cassandra assigns the following hash values to each partition key:

Partition Key	Murmur3 Hash Value
Jim	-2245462676723223822
Carol	7723358927203680754
Johnny	-6723372854036780875
Suzy	1168604627387940318

Cassandra determines the nodes on which it ought to store the four values according to the value of the partition key. As you can see, the partition key is hashed to a specific range and each node is assigned a range. In this example, Cassandra stores the four rows in the four nodes as shown here:

Node	Start Range	End Range	Partition Key	Hash Value
A	-9223372036854775808	-4611686018427387904	Johnny	-6723372854036780875
B	-4611686018427387903	-1	Jim	-245462676723223822
C	0	4611686018427387903	Suzy	1168604627387940318
D	4611686018427387904	9223372036854775807	Carol	7723358927203680754

Note   You must use the same token architecture for all nodes in a datacenter, whether it be a single-token or a vnode-based architecture.

How to Distribute Token Ranges

When you enable vnodes, there are two ways to distribute token ranges in a datacenter: the *allocation algorithm* and the *random selection algorithm*. I explain these algorithms in this section.

The Allocation Algorithm

The allocation algorithm, which is supported only for the two partitioners Murmur3Partitioner and RandomPartitioner, strives to balance workload using few tokens. You can use a large number of tokens to spread the workload around, but it means you must also manage all those tokens.

The allocation algorithm uses the num_tokens property setting to distribute the token ranges proportionally. For a given replication factor, this algorithm tries to choose tokens so it optimizes the replicated load across the cluster's nodes. It assigns each node a load proportional to the number of nodes.

Tip   The recommended practice is to use 8 vnodes (tokens) to provide an ideal balance between efficient distribution of the workload and performance. The allocation algorithm optimally distributes token ranges among the nodes and racks by using the keyspace replication factor of the datacenter. If you're using Cassandra version 3.4 or lower, you need to use either 256 nodes or create and destroy the vnodes until you arrive at a satisfactory distribution of data.

To set the allocation algorithm, configure the `allocate_tokens_for_local_` `replication_factor` property to the replication factor of the keyspaces in a datacenter.

The Random Selection Algorithm

The random selection algorithm will randomly distribute token ranges across a datacenter's nodes. To use this algorithm, you must configure the `num_tokens` property.

The database distributes the token ranges to a node based on the number of tokens it assigned to the nodes in a datacenter. For example, you can set the number of vnodes (tokens) to 128 for a transaction-heavy datacenter.

Enabling Vnodes

Unless the hardware is different on the nodes of a cluster, you should configure the same number of vnodes on each node. If the hardware capacities differ among the nodes, you can assign a different number of vnodes to nodes, such as 128 vnodes for smaller machines and 256 vnodes for machines that are much more powerful. While this is technically possible, it's not a recommended practice because you're likely to end up with unbalanced datasets.

Cassandra uses vnodes by default. You set the number of the tokens on a node with the `num_tokens` parameter. The default value for this parameter is 256. If you decide not to use vnodes and configure traditional token ranges instead, you must disable vnodes by setting the `num_tokens` property to 1 or by commuting the property. You must then set the `initial_token` property in each node to specify the token ranges that'll be owned by that node.

When you configure vnodes with the `num_tokens` parameter, don't set the `initial_token` parameter; this parameter is if you want to specify a single token, as explained earlier.

Tip   The `initial_token` parameter is only for legacy clusters. All new clusters should use the `num_tokens` parameter to configure vnodes. The `initial_token` parameter allows you to specify tokens manually.

Disabling Vnodes

If you decide not to use vnodes, you must ensure that each node is assigned approximately the same amount of data. You can do this by assigning an initial token value for each node and then calculating the tokens for each datacenter.

To disable vnodes, configure the `cassandra.yaml` file as follows:

1. Comment both the `num_tokens` and the `allocate_tokens_for_local_replication_factor` properties.

2. Set the `initial_token` property to 1.

Choosing a Replication Strategy

Replication is how Cassandra stores copies of data on multiple nodes for high availability and fault tolerance. The replication factor determines how many replicas Cassandra will store for each row in a table. A replication factor of 1 means Cassandra will keep a single copy of each row, and a replication factor of 2 means that the database will store two copies of each row, with the two copies stored on two different nodes.

Cassandra stores replicas on multiple nodes for reliable and fault tolerant operations. The term *replication strategy* refers to how Casandra determines precisely on how many nodes it ought to store the replicas. If your cluster has several nodes, you can set a replication factor higher than 1 to ensure that the table data is available even if one or more nodes in the cluster are unavailable.

The database will store the first replica of data on the node that owns the range in which the token falls. It uses the replication strategy you configure to determine on which nodes it should place the rest of the replicas.

Obviously, specifying a replication factor of 1 means that if the node with a replica crashes, you can't access that row. Also, the higher the replication factor, the more storage you need to store the rows.

A *replication group* determines how many replicas you store in a single datacenter. You don't need to store the same number of replicas in all datacenters in a cluster. For example, you can store three replicas in a datacenter to service real-time applications and a single replica in a different datacenter to serve analytical queries.

There are two basic replication strategies:

- SimpleStrategy
- NetworkTopologyStrategy

The following sections explain the two basic replication strategies.

SimpleStrategy

Under SimpleStrategy, Cassandra stores the first replica on a node that the partitioner determines and places the rest of the replicas on the remaining nodes clockwise in the ring. This strategy ignores the cluster topology, which refers to the location of the racks and datacenters.

Tip   For most deployments, NetworkTopologyStrategy is highly recommended because it makes it easier to expand to multiple datacenters in the future.

The SimpleStrategy replication strategy is good only if you have a single datacenter and rack. You must use the NetworkTopologyStrategy if you're planning to use more than one datacenter.

NetworkTopologyStrategy

Under NetworkTopologyStrategy, you specify how many replicas you want to have in each datacenter. So obviously, this strategy is designed for multiple datacenter clusters.

NetworkTopologyStrategy tries to place replicas on multiple racks since all nodes in a rack tend to fail at the same time. When using this strategy, your main concern should be the tradeoff between local reads to reduce latency and the possibility of potential failures.

Most common configurations when dealing with multiple datacenters are the following:

- *Two Replicas per Datacenter*: This strategy will let you survive the failure of a single node in each replication group while enabling local reads (consistency level of ONE).

- *Three Replicas per Datacenter*: This strategy will let you survive the failure of multiple nodes in each datacenter with a consistency level of ONE. It also helps you tolerate the failure of a single node per replication group while using the LOCAL_QUORUM consistency level,

Dynamic Ring Participation

You can start and stop nodes in a Cassandra cluster without any effect on the rest of the nodes. In the following example, you shut down one of the two nodes in the test cluster. You then check the log of the second node.

```
INFO  [GossipTasks:1] 2017-06-25 08:35:01,697 Gossiper.java:1035 -
InetAddress /192.168.159.129 is now DOWN
```

You then start up the node you brought down earlier. You again check the logs of the good node.

```
INFO  [HANDSHAKE-/192.168.159.129] 2017-06-25 08:39:05,030
OutboundTcpConnection.java:510 - Handshaking version with /192.168.159.129
INFO  [GossipStage:1] 2017-06-25 08:39:06,134 Gossiper.java:1054 - Node
/192.168.159.129 has restarted, now UP
INFO  [GossipStage:1] 2017-06-25 08:39:06,139 StorageService.java:2248 -
Node /192.168.159.129 state jump to NORMAL
INFO  [GossipStage:1] 2017-06-25 08:39:06,170 TokenMetadata.java:479 -
Updating topology for /192.168.159.129
INFO  [RequestResponseStage-1] 2017-06-25 08:39:06,352 Gossiper.java:1020 -
InetAddress /192.168.159.129 is now UP
```

As you can see, Cassandra automatically detects the failure and the restart of nodes. You don't need do a thing.

Changing the Replication Strategy

You set the replication factor and the replication strategy at the keyspace level. The following example shows how to switch a keyspace from the default `SimpleStrategy` to `NetworkTopologyStrategy`:

```
cqlsh> ALTER KEYSPACE cycling WITH REPLICATION = {'class' :
'NetworkTopologyStrategy', 'DC1' : 3};
```

In this example, you switch the keyspace replication strategy for a single datacenter named DC1.

In the following example, you restrict the replication of a keyspace to selected datacenters and set the replication factor of the excluded datacenters to 0:

```
cqlsh> ALTER KEYSPACE cycling WITH REPLICATION = {'class' :
'NetworkTopologyStrategy', 'DC1'  : 0, 'DC2' : 3, 'DC3' : 0 };
```

Partitioners and Partitioning Strategies

Cassandra uses partitioners to help it distribute data across the nodes in a cluster. A partitioner load balances the data by distributing it evenly across all nodes in a cluster. A partitioner is a function that Cassandra uses for deriving tokens that represent a row by hashing the primary keys. Cassandra then distributes the data across the nodes based on the token values it derives through the partitioner.

Types of Partitioners

Cassandra offers three types of partitioners, although it recommends that you not use the `ByteOrderedPartitioner` due to various drawbacks and includes it only for backward compatibility purposes.

The other two partitioners that you can choose from are the `Murmur3Partitioner`, which is the default partitioner, and the `RandomPartitioner`. Both use tokens to assign equal amounts of data on all nodes and evenly distribute table data across the ring or keyspace.

Since each part of the hash range contains an equal number of rows (on average), read and write requests are evenly distributed.

Both of the partitioners distribute data across the cluster based on the hash values for each of the row keys. The difference between the two partitioners is in how each of them generates the token hash values. The Murmur3Partitioner provides faster hashing using the Murmurhash function, which creates a 64-bit hash value of the partition key.

The RandomPartitioner distributes data across the cluster by generating token hash values using a cryptographic hash. Generation of the token hash values takes longer with the RandomPartitioner due to its use of a cryptographic hash. Since Cassandra doesn't need a cryptographic hash, using the alternative, Murmur3Partitioner, will yield a 3-5 times performance gain.

Tip   The Murmur3Partitioner is 3-5 times faster in performance than the RandomPartitioner.

The Murmur3Partitioner, which is the default partitioning strategy, is good in most cases. Just be aware that once you partition data with one partitioner, it's difficult to convert it to the other partitioner.

Generating Tokens

If you're using virtual nodes (vnodes), there's no need to calculate tokens. Otherwise, you must calculate tokens for the cluster. You assign these tokens with the initial_token parameter when configuring the cluster (cassandra.yaml file).

If you have just one datacenter, you calculate the tokens by dividing the hash range by the number of nodes. If you have multiple datacenters, you must calculate tokens for each datacenter by dividing the hash range by the number of nodes in each datacenter.

How you calculate tokens depends on your partitioning strategy. Assuming you're using the default Murmur3Partitioner, you generate tokens in the following way:

```
python -c "print [str(((2**64 / number_of_tokens) * i) - 2**63) for i in range(number_of_tokens)]"
```

If you have six nodes in a cluster, this is the Python command you run:

```
python -c "print [str(((2**64 / 6) * i) - 2**63) for i in range(6)]"
```

The command will display tokens for each of the six nodes in the cluster, which you can then specify as the value of the initial_token parameter in the cassandra.yaml file.

```
[ '-9223372036854775808', '-6148914691236517206', '-3074457345618258604',
'-2', '3074457345618258600', '6148914691236517202' ]
```

Setting the initial_tokens and num_tokens Properties

By default, the `initial_token` property is disabled for a single-node-per-token architecture. Where a node owns one contiguous range in the ring space, you must set the `initial_token` property. When you set this, it'll override the `num_token` property.

Always set the `initial_token` value when initiating a production cluster for the first time or when adding nodes, if any of the following is true:

- You're not using vnodes.

- You've set the node's `num_tokens` property to 1.

You set the `num_tokens` property for a virtual node architecture to determine the number of token ranges to assign to a vnode. The default value for this property is 1 disabled, which means vnodes are disabled. You can specify a number between 1 and 256 for this property.

If all nodes in the cluster are physically identical, you must specify the same value for the `num_values` property on all the nodes.

Snitches

A snitch's job is to determine where each node is, relative to the rest of the nodes in a cluster. Snitches help the database identify the fastest way to respond to queries. Snitches help Cassandra figure out which datacenters and racks it should use to read data from and write data to. Snitches determine how Cassandra distributes replicas. Snitches make Cassandra aware of the network topology so the database can route requests efficiently by grouping nodes into datacenters and racks.

The snitch calculates the distance between the nodes and finds out which nodes lie close to each other. It uses this information to help the database route requests to the best nodes. The best node here is the replica that will return the data the fastest.

A snitch is a critical component of Cassandra's architecture and helps determine the datacenter and rack to which a node belongs. A snitch maps the IP addresses of nodes in a cluster to racks and datacenters. A rack is a physical entity and a data center is a virtual entity.

Snitches are quite critical to read activity. When reading data, Cassandra asks only one node for the data and, subject to the consistency level and the read repair chance you've configured, requests only checksums from the other replicas.

When there are multiple replicas from which the coordinator node can make a request for the actual data, Cassandra needs to make a choice as to which replica it should ask to send in the full data (not just checksums). The snitch monitors the read performance of the various replicas and chooses the best replica based on the historical performance data of the replicas.

Note   The choice of a snitch affects where Cassandra places replicas. The purpose of a snitch is to route requests efficiently and to distribute replicas evenly.

A key goal of Cassandra is to avoid storing multiple replicas of data on the same rack. The replication strategy you adopt uses the information provided by the new snitch to place the replicas.

All snitches are dynamic by default, which enables Cassandra to move requests away from nodes that are currently having performance issues. You can configure dynamic snitch thresholds for a node in the cassandra.yaml file. As the performance of a "bad" node improves, it can get back to a preferred status.

Types of Snitches

Cassandra offers several types of snitches, as explained in the following sections.

- SimpleSnitch: This is the default snitch, but it is good only for development environments. This snitch is unaware of datacenters or racks and thus is unusable for multi-datacenter environments.

- GossipingPropertyFileSnitch: This snitch relays the rack and datacenter information you configure in the cassandra-rackdc. properties file to the rest of the nodes using gossip.

You can configure the GossipingPropertyFileSnitch for a node by editing its cassandra-rackdc.properties file as shown here:

```
dc=DC1
rack=RACK1
prefer_local=true
```

Tip   DataStax recommends GossipingPropertyFileSnitch for production usage.

Here, dc refers to datacenter and rack specifies the rack information. The prefer_local option specifies that Cassandra must use the local IP address when it's not communicating across multiple datacenters in order to limit the network bandwidth usage.

- Ec2Snitch: This is a simple snitch for Amazon EC2 deployments where all nodes are in a single region. The region name is analogous to the datacenter name, and the availability zones are analogous to racks within the datacenter.

- Ec2MultiRegionSnitch: You use this snitch for Amazon EC2-based clusters where the clusters span multiple regions.

- GoogleCloudSnitch: This is the snitch for a Cassandra deployment on the Google Cloud Platform across a single or multiple regions.

- RackInferringSnitch: This snitch figures out the node location by rack and datacenter. In a node IP address, which has four octets (110.100.200.105, for example), the third and fourth octets correspond to the rack and datacenter. This snitch is useful for writing custom snitch classes.

- PropertyFileSnitch: Determines the closeness of the nodes based on the rack and datacenter they belong to by using the network definitions from the cassandra-topology.properties file. You must define all nodes in the cassandra-topology.properties file, and the file must be identical on all nodes of a cluster.

- CloudstackSnitch: This is a snitch for an Apache Cloudstack-based cluster.

Note   Snitches inform Cassandra about the network topology so the database can efficiently route the requests and evenly distribute replicas.

Understanding the cassandra-topology.properties and the cassandra-rackdc.properties Files

The `cassandra-topology.properties` and the `cassandra-rackdc.properties` files play an important role in configuring snitches, so I briefly explain these two files here.

The `cassandra-topology.properties` file contains the cluster topology for the entire cluster. The following are the contents of an example `cassandra-topology.properties` file:

```
# datacenter One

175.56.12.105=DC1:RAC1
175.50.13.200=DC1:RAC1
175.54.35.197=DC1:RAC1

120.53.24.101=DC1:RAC2
120.55.16.200=DC1:RAC2
120.57.102.103=DC1:RAC2

# datacenter Two

110.56.12.120=DC2:RAC1
110.50.13.201=DC2:RAC1
110.54.35.184=DC2:RAC1

50.33.23.120=DC2:RAC2
50.45.14.220=DC2:RAC2
50.17.10.203=DC2:RAC2
```

This example shows a file with two physical datacenters, each with two racks. The `PropertyFileSnitch` uses the `cassandra-topologies.properties` file. If you don't identify any of a cluster's nodes in the `cassandra-topologies.properties` file, the database assumes that they are in the default datacenter (datacenter) and rack (rack1).

You must update this file as you add and delete nodes to your cluster in order to make Cassandra aware of the rack and datacenter the nodes belong to. While Cassandra can figure these things out on its own, from a performance viewpoint it's better for you to feed that information to Cassandra.

Here are the contents of a typical `cassandra-rackdc.properties` file:

```
# indicates the rack and dc for this node
dc=DC1
rack=RAC1
```

When you configure the `GossipingPropertyFileSnitch`, it always loads the `cassandra-rackdc.properties` file, should there be one.

The following three snitches will look up the `cassandra-rackdc.properties` file to figure out to which datacenter and racks the nodes in the cluster belong:

- `GossipingPropertyFileSnitch`

- `Ec2Snitch`

- `Ec2MultiRegionSnitch`

Summary

Understanding how Cassandra stores data and how it performs read and write operations is critical to tuning database performance.

Repairing Cassandra nodes to correct data inconsistency is an essential task, and it's good to master the various types of manual repair that Cassandra offers.

Understanding how to configure the `cassandra-topology.properties` and `cassandra-rackdc.properties` files helps configure data centers and clusters.

Introduction to the Cassandra Query Language

This chapter provides a quick summary of the Cassandra Query Language (CQL).

Understanding how to perform various types of DML (data manipulation language) operations such as selecting, inserting, and updating data is critical, and you'll learn the intricacies of Cassandra's DML operations. Understanding DDL operations such as creating and removing objects is important, and this chapter explains how to perform various types of DDL operations.

The chapter shows you how to create structures such as keyspaces and tables. You'll learn how to create primary keys and you'll examine the role of secondary indexes in Cassandra. Materialized views are a better alternative to secondary indexes and you'll learn how to work with them.

I review the various CQL data types including advanced data types such as collections, types, and user-defined types (UDTs).

The chapter also explains user-defined functions (UDFs) and user-defined aggregates (UDAs).

Working with Keyspaces

Cassandra uses keyspaces, which are logical entities, to group tables. A keyspace is somewhat like a named database that you use in a relational database system. However, the real purpose of a keyspace is to act as a namespace that specifies how Cassandra replicates data. This means that if you have different sets of data that differ in their replication requirements, you can use different keyspaces to store the data.

189

© Sam R. Alapati 2018
S. R. Alapati, *Expert Apache Cassandra Administration*, https://doi.org/10.1007/978-1-4842-3126-5_6

A keyspace is a logical structure where Cassandra stores not only table data, but also all other entities that you create for an application, such as materialized views, functions, aggregates, and UDTs.

Note   You control Cassandra's data replication on a per-keyspace basis.

In the following sections, I show how to manage keyspaces.

Managing Keyspaces

Cassandra stores its data in tables and it groups the tables into *keyspaces.* A keyspace is a logical container, and it lets you define options that apply to all the tables in that keyspace.

Note   Cassandra recommends that you use a single keyspace for each application.

When you refer to a table, you must fully qualify it by providing the keyspace in which that table lives. You don't need to provide the keyspace name if the table belongs to the current keyspace. As you'll learn shortly, you specify the current tablespace by executing the USE statement.

Creating a Keyspace

Before creating a keyspace, you must log into cqlsh. Once you do so, run the `describe cluster` command to view details about your cluster.

```
$ cqlsh
Connected to Test Cluster at 127.0.0.1:9042.
[cqlsh 5.0.1 | Cassandra 3.10 | CQL spec 3.4.4 | Native protocol v4]
Use HELP for help.
cqlsh> describe cluster;
Cluster: Test Cluster
Partitioner: Murmur3Partitioner
cqlsh>
```

You create a keyspace with the CREATE KEYSPACE SQL statement. Here's the syntax of the statement:

```
CREATE  KEYSPACE [IF NOT EXISTS] keyspace_name
   WITH REPLICATION = {replication_map}
   [AND DURABLE_WRITES =  true|false] ;
```

You can specify two options when creating a keyspace: the first is replication and the second is durable_writes. Here's what the two options mean:

- replication: This is a mandatory option that specifies the replica placement strategy and the number of replicas. The replication_map attribute allows you to specify then number of copies of the data in a datacenter.

  The replication option must contain the sub-option 'class', which specifies the replication strategy. A replication strategy tells Cassandra where it should store copies of the data in this keyspace across the cluster's datacenters and racks. There are two possible replication strategies:

  - SimpleStrategy: As its name indicates, this uses a simple replication strategy by setting the same replication factor cluster-wide for all data centers within the cluster. Here's an example:

    ```
    {'class': 'SimpleStrategy', 'replication_factor' : 3};
    ```

    If you are running a test cluster with just a single node, your replication factor must be 1. SimpleStrategy is only good for development and proof-of-concept uses. For production purposes, you must use NetworkTopologyStrategy.

Tip   Never use a replication factor of 1 for storing data that you can't afford to lose!

  - NetworkTopologyStrategy: Lets you specify a more complex replication strategy by allowing you to set the replication factor separately for each data center. Example:

    ```
    {'class': 'NetworkTopologyStrategy', 'DC1' : 1, 'DC2' : 3}
    ```

You must specify a datacenter name in the snitch properties file or use a single datacenter named datacenter1 to use NetworkTopologyStrategy.

The second part of the replication option allows you to set the number of copies of data in this keyspace by configuring a value for the replication_factor attribute.

- durable_writes: Tells Cassandra whether it ought to use the commit log for any updates in the current keyspace. The default value is true and you should use this value in production databases.

   You can bypass the writing of changes to the commit log first by specifying false as the value for durable_writes. This will speed up the writes to the tables in this keyspace, but should the node go down before the database flushes the memtables to the SSTables, you risk losing the data.

Caution   Don't disable durable-writes when you configure SimpleStrategy for replication.

Here's an example that shows how to create a keyspace named *cycling*:

```
cqlsh> CREATE KEYSPACE IF NOT EXISTS cycling
      WITH REPLICATION = { 'class' : 'NetworkTopologyStrategy',
      'datacenter1' : 3 };
cqlsh>
```

If there are no errors, you can be sure that the database created the keyspace you want. You can run the describe keyspaces command to ascertain that the database did indeed create the keyspace.

```
cqlsh> describe keyspaces;
cycling  system_schema  system_auth  system  system_distributed  system_
traces
cqlsh>
```

As you can tell, the IF NOT EXISTS part of the CREATE KEYSPACE statement is optional and helps avoid an error should the keyspace *cycling* already exist.

Note A keyspace is the defining container for replication.

You can see the new keyspace *cycling* in the list of keyspaces shown by the database.

Next, you can execute the describe cycling command to view details about this keyspace.

```
cqlsh> describe cycling;
CREATE KEYSPACE cycling WITH replication = {'class':
'NetworkTopologyStrategy', 'datacenter1': '3'}  AND durable_writes = true;
cqlsh>
```

Cassandra stores the history of all your nodetool and cqlsh commands in separate files in the $CASSANDRA_USER/.cassandra directory, with the user being the one that started the cluster.

```
$ ls -altr
drwx------ 6 root root 4096 May 26 07:25 ..
-rw-r--r-- 1 root root  127 Jun 23 10:13 nodetool.history
-rw------- 1 root root 1308 Jun 24 07:24 cqlsh_history
drwxr-xr-x 2 root root 4096 Jun 24 07:24 .
$
```

Creating a Keyspace in a Cluster with Multiple Datacenters

You can create a keyspace with a different replication factor for each of the datacenters, as shown in the following example:

```
cqlsh> CREATE KEYSPACE "Cycling"
       WITH REPLICATION = {
       'class' : 'NetworkTopologyStrategy',
       'datacenter1'  : 3 ,
       'datacenter2 ',  2 ,
       'datacenter3' :, 1 ,
};
```

Note   Objects such as tables, functions, and UDTs are bound to a specific keyspace. To work with the objects, you must either specify the keyspace name each time you access an object or make that keyspace your current keyspace.

Altering a Keyspace

Occasionally you may want to alter a keyspace, meaning you want to change one of the two properties that you can set when creating a keyspace. These two properties, which I described in the "Creating a Keyspace" section, are replication and durable_writes.

Changing the Replication Factor

If you want to change the replication factor of a keyspace, you can do it by executing the ALTER KEYSPACE command, which has the following syntax:

```
ALTER KEYSPACE "KeySpace Name"
WITH replication = {'class': 'Strategy name', 'replication_factor' : 'No.Of
replicas'};
```

Note   You set the replication strategy at the keyspace level when creating the keyspace or later by modifying the keyspace.

Here's an example that shows how to change the replication from 1 to 3, by altering a keyspace named *cycling*:

```
cqlsh> ALTER KEYSPACE cycling
     WITH replication = {'class':'SimpleStrategy',
     'replication_factor': 3};
```

Note   You can't alter the name of a keyspace.

If you've configured NetworkTopologyStrategy, you need to do more. Since the change in the replication factor affects all the nodes that the keyspace will replicate to, or not replicate to, you must also prepare the affected nodes for the change in the replication factor. Here are the procedures for changing the replication factor in this situation:

1. Change the replication level in all datacenters.

    ```
    cqlsh> ALTER KEYSPACE cycling WITH REPLICATION =
           {'class' : 'NetworkTopologyStrategy', 'dc1' : 3,
           'dc2' :2};
    ```

2. Run the nodetool repair command as shown here to perform a full repair of the keyspace:

    ```
    $ nodetool repair -full
    ```

Once the repair completes on a node, start the repair on the next node.

The Need for Running the nodetool repair Command

When you raise the replication factor, the change doesn't automatically take effect on all nodes. After you raise the replication factor for a cluster, you must run the nodetool repair command on all the nodes in that cluster. Similarly, you must run the nodetool repair command on all nodes in a data center when you raise the replication factor for a data center.

Running the nodetool repair command tells Cassandra to create the additional replicas it needs to satisfy the replication factor you configure.

Tip   The write throughput of the cluster is inversely related to the replication factor.

If you lower the replication factor, you must run the nodetool cleanup command afterwards, whether the change is in a datacenter or in the cluster. This will enable Cassandra to free up space occupied by the replicas it doesn't need any longer.

Preventing a Keyspace from Sending Replicas to Some Data Centers

Sometimes you may want to prevent a keyspace from writing replicas only in some datacenters, or allow a keyspace to write to just one specific datacenter. By setting the replication factor to zero (0), you can keep a keyspace from sending replicas to a specific datacenter. The following is an example:

```
cqlsh> ALTER KEYSPACE cycling
       WITH REPLICATION = {'class' : 'NetworkTopologyStrategy', 'dc1' :3,,
       'dc2': 0, 'dc3': 3 };
```

This command excludes the datacenter DC2 (by setting the replication factor to 0).

Altering the Durable Writes Property

The ALTER KEYSPACE command also lets you change the durable_writes property. The following is an example that shows how to change the durable_writes property of a keyspace:

```
cqlsh> ALTER KEYSPACE cycling
       WITH REPLICATION = {'class' : 'NetworkTopologyStrategy',
'datacenter1'
       : 3}
       AND DURABLE_WRITES = true;
```

You can confirm that the changes you've made have stuck by running the following command.

```
cqlsh> SELECT * FROM system.schema_keyspaces;
keyspace_name | durable_writes | strategy_class | strategy_options
----------------+----------------+------------------------------------------
---
cycling | True | org.apache.cassandra.locator.NetworkTopologyStrategy |
{"datacenter1":"3"}
mykeyspace | True | org.apache.cassandra.locator.SimpleStrategy |
"replication_factor":"4"}
...
```

Repairing a Keyspace

When you add a datacenter, be sure to run the following `nodetool repair` command to perform a full repair of the keyspace:

```
$ nodetool repair -full cycling;
```

Specifying the Keyspace You Want to Use

Often, you're working within a certain keyspace and you want to move to a different keyspace and make the new keyspace your current or working keyspace. You must specify the keyword USE (what else?) to make a keyspace your current default keyspace.

In the following example, you want a keyspace named *myKeyspace* to be the default keyspace:

```
cqlsh> USE myKeySpace;
cqlsh:myKeySpace>
```

Using the Keyspace Qualifier

When your code deals with multiple keyspaces, it becomes cumbersome to issue the USE <KEYSPACE> command. In these situations, you can simply specify the *keyspace qualifier* instead of having to execute the USE <KEYSPACE> command. Cassandra allows you to specify the keyspace qualifier when executing the following statements:

- ALTER TABLE
- CREATE TABLE
- DELETE
- INSERT
- SELECT
- TRUNCATE
- UPDATE

To specify a table in a keyspace, specify the name of the keyspace followed by a period and then the table name. The following example shows how to insert data into a table named race_winners, which is in the *cycling* tablespace.

```
cqlsh> INSERT INTO cycling.race_winners (race_name, race_position,
cyclist_name) VALUES (...);
```

Removing a Keyspace

Removing a keyspace is easy. Execute the DROP KEYSPACE statement to remove a keyspace, as shown here:

```
cqlsh> DROP KEYSPACE cycling;
cqlsh>
```

As you can see, when you issue the DROP KEYSPACE statement, Cassandra doesn't say anything; it does drop the keyspace, and you can verify that the keyspace is gone by running the DESCRIBE KEYSPACES command.

```
cqlsh> DESCRIBE KEYSPACES;
system_traces  system_schema  system_auth  system  system_distributed
cqlsh>
```

When you remove a keyspace, the database removes all its constituent entities such as tables, aggregates, types, and UDFs. Cassandra can automatically back up the keyspace if you've set the auto_snapshot property in the cassandra.yaml file.

By default, the auto_snapshot property has the value true, which means that when you drop a tablespace (or a table), the database automatically takes a snapshot of the keyspace (or table). You also have the option of manually backing up a keyspace before dropping it, as explained in Chapter 9.

Note   DataStax strongly recommends that you leave the default value of true intact for the auto_snapshot property.

System Keyspaces

Cassandra uses a set of *system keyspaces* to store details about the configuration of the cluster and the objects that are stored in that keyspace.

The DESCRIBE KEYSPACES command shows all the system keyspaces in a cluster.

```
cqlsh> DESCRIBE KEYSPACES;
system_traces system_schema  system_auth system  system_distributed
cqlsh>
```

Tip   By default, the replication factor for the *system-auth* keyspace is set to 1. DataStax recommends that you set the replication factor for the *system-auth* tablespace to the number of nodes in each of the datacenters. However, you can set the replication factor for this keyspace to 5 or 7, which offers plenty of redundancy.

The following is a selective description of data stored in various system keyspaces:

- *system*: This keyspace contains tables that contain information pertaining to materialized views, hints, indexes, peers, information on partitions, and prepared statements used by drivers.

- *system_schema*: This keyspace contains information about table columns and indexes, UDFs, triggers, user-defined types, and materialized views.

- *system_distributed*: This keyspace has a single table named `repair_history`, which stores information about keyspace repair activity.

Getting Cluster Topology from the System Tables

You can get the cluster topology information such as the IP address of peer nodes, the names of the datacenter and racks, and token values by querying the system table PEERS, as shown here:

```
cqlsh> SELECT * FROM system.peers;
 peer            | data_center | host_id                              | preferred_ip
 | rack  | release_version | rpc_address    | schema_version
                      | tokens
192.168.159.129 |         dc1 | 632fe5cd-26fa-4b8b-9842-182be2f954e5
 |        null | rack1 |            3.10 | 192.168.159.129 | 86afa796-d883
 '-1470738432828591905', '-1533929660503050360', '-1588268933422992241',
 '-1790127126337716384', '-1795704774997895089', '-1933674708843479324',
 '-2001904010895819169', '-2137322678046758761', '-2315125955255423142',
 '-2349800311170714405', '-2390011164349178402', '-2439785688421178687', '-
 ...
```

Getting Information About Functions, Aggregates, and User Types

Later on in this chapter, you'll learn about Cassandra's user-defined entities such as UDFs, aggregates, and types. Once you create the user-defined entities, you can use the system tables to retrieve information about these entities.

To retrieve information about UDFs, query the system table SCHEMA_FUNCTIONS.

```
cqlsh> SELECT * FROM system.schema_functions;
```

You can get details about the user-defined aggregates by querying the SCHEMA_AGGREGATES table.

```
cqlsh> SELECT * FROM system.schema_aggregates;
```

Finally, you can query the SCHEMA_USERTYPES table to view all user-defined types in the database.

```
cqlsh> SELECT * FROM system.schema_usertypes;
```

Now that you know how to work with keyspaces, it's time to go ahead and use them for what they're meant for: to act as logical repositories for tables and other entities by creating a table.

Creating a Table

A Cassandra CQL table stores rows and it has a name. When you create a table, you define the columns for the rows, a mandatory *primary key* to identify each row, and any additional options you may choose.

You must execute the CREATE TABLE statement to create a new table. The core syntax for the CREATE TABLE statement is as follows:

```
create_table_statement ::=  CREATE TABLE [ IF NOT EXISTS ] table_name
                            '('
                            column_definition
                            ( ',' column_definition )*
                            [ ',' PRIMARY KEY '(' primary_key ')' ]
                            ')' [ WITH table_options ]
```

The following is a typical table creation statement:

```
cqlsh> use cycling;
cqlsh:cycling> CREATE TABLE loads (
        ...        machine inet,
        ...        cpu int,
        ...        mtime timeuuid,
        ...        load float,
        ...        PRIMARY KEY ((machine, cpu), mtime)
        ...        ) WITH CLUSTERING ORDER BY (mtime DESC);
cqlsh:cycling>
```

Before you create the table loads, you run the command use cycling so the database will know to create the new table in this keyspace. Alternatively, you could have specified CREATE TABLE cycling.loads. This will also let the database know in which keyspace you want it to place your new table.

In the table creation statement, the following are the main things you need to focus on:

- Column definitions

- Primary and partition keys

- Clustering columns

- Table options (compact storage and clustering order)

In the following sections, I briefly review the key entities that are part of a table creation statement.

Note   A primary key consists of two things: the first column or columns is the mandatory *partition key*, followed by one or more *clustering columns*.

Column_definition

A `column_definition` clause consists of the name of the column and its type, as well as two modifiers.

- *Static*: Declares the column as a static column. A static column has the same value for all rows that share the same partition key (explained in a little bit). Of course, only non-primary keys can be static.

- *Primary key*: A primary key uniquely identifies a row, and all tables must define a primary key. I explain primary keys and partition keys in the following section.

Primary Keys, Partition Keys, and Clustering Columns

Cassandra's concept of a primary key is quite different from that of a normalized relational database. The primary key of a table specifies the location and the order of data stored in that table.

When you define a primary key, it can have two parts: a *partition key* and a *clustering key*. At a minimum, it must have a partition key. Here's what the two parts mean:

- Cassandra uses the first part of the definition of a primary key, the partition key, to distribute the data in the table across the cluster's nodes. The partition key determines which node will store a specific row of the table. A compound partition key can split the data to store related data on separate partitions.

- The database uses the second part of the key definition, called the clustering key or *clustering column* (or columns), to order or sort the data within the partition.

You can specify multiple columns when defining the partition key and clustering key of a primary key. The resulting key is a *composite primary key*. The idea behind this type of a design is to distribute table data easily throughout the nodes of a cluster. It also allows for higher performance and facilitates failover.

Tip   Once you create a table, you can't change its primary key. You must create a new table and insert data into the new table instead.

Choosing the right primary keys, partition keys, and clustering columns is a key aspect of Cassandra data modeling, with the choice significantly affecting query performance.

A primary key is mandatory for a table. You can specify a primary key with one or more columns of a table, and the order of the columns matters. A primary key on a Cassandra key is unique in the sense that it contains more elements than a typical primary key in a relational database.

As mentioned, a primary key consists of two parts: the partition key and the clustering columns. You'll learn more about these two components of a primary key in the following sections.

Note   A partition key groups rows in the same replica set. The clustering columns dictate how rows are stored in the replica.

The Partition Key

The partition key is the first and mandatory component of a primary key. It can include one or more columns. Here's an example:

```
CREATE TABLE t (k text PRIMARY KEY)
```

Here, column k is the partition key, and there are no clustering columns.

A table partition is a set of rows that have the same value for their partition key. A partition key can have one or more columns, and if it's a multi-column key, the values for all the partition key columns must be identical for the rows in a partition.

Here's an example that illustrates the concept of the partition key:

```
CREATE TABLE t (
    a int,
    b int,
    c int,
    d int,
    PRIMARY KEY ((a, b), c, d)
);
```

Let's say you execute the following simple query on this table:

```
SELECT * FROM t;
   a | b | c | d
  ---+---+---+---
   0 | 0 | 0 | 0      // row 1
   0 | 0 | 1 | 1      // row 2
   0 | 1 | 2 | 2      // row 3
   0 | 1 | 3 | 3      // row 4
   1 | 1 | 4 | 4      // row 5
```

In this case, note the following:

- Row 1 and row 2 share the same partition.

- Row 3 and row 4 share the same partition.

- Row 5 is in a different partition.

Note   A primary key is the same as the partition key when there are no clustering columns.

Here are the key properties of a partition:

- Cassandra guarantees that all rows from a partition are stored in the same set of replica nodes. You should therefore choose the partition key carefully, so you can place rows that you often query together in the same partition. This means that the database must do less work in searching for the data on multiple nodes.

Tip   If a partition key contains too much data, it can lead to the creation of a "hot spot."

- Cassandra performs all updates in a single partition atomically and in isolation.

Creating a Table with a Simple Primary Key

You specify a single column name as the partition key when you define a table with a simple primary key. Thus, the primary key and the partition key are the same. If you have a column with numerous values, it is easy to insert and query the data since Cassandra distributes the partitions across multiple nodes.

If your application requires a simple lookup table with a single identifier, then a simple primary key is the way to go. In the following example, you specify id as the primary key, which makes it easy to get the names of cyclists by providing their ID numbers;

```
cqlsh> USE cycling;
cqlsh> CREATE TABLE cyclist_name ( id UUID PRIMARY KEY, lastname text,
firstname text );
```

The column id is in the UUID format. UUID (universally unique identifier) is a 128-bit value and the bits conform to various types. In CQL, the uuid type is a Type 4 UUID, based on random numbers. An UUID is usually represented by a sequence of hex digits separated by dashes.

It's common to use the uuid type as a surrogate key, either by itself or by combining it with other values. You can generate a Type 4 UUID value and use that value in an INSERT or UPDATE statement, as shown here:

```
cqlsh:cycling> insert into cyclist_name (id, firstname, lastname)
            ... values
            ... (uuid(), 'sam', 'alapati');
cqlsh:cycling> select * from cyclist_name;
 id                                   | firstname | lastname
--------------------------------------+-----------+----------
 3c4acfde-dc65-4c9f-b2ec-9047d677641a |       sam |  alapati

(1 rows)
cqlsh:cycling>
```

Tip   Another way to generate a UUID is to use the built-in CQL function NOW. For example, you can do the following to generate the UUID from the current time:

```
cql> INSERT INTO "users" ("username, id, 'address" VALUES ('alapati',
NOW(), '12345, Main St, Anytown');
```

Using the NOW function lets you generate a unique UUID for each row you insert into the USERS table, without having to explicitly specify a UUID constant.

You can also specify the primary key at the end of the table definition, as shown here:

```
cqlsh> USE cycling;
cqlsh> CREATE TABLE cyclist_name ( id UUID, lastname text, firstname text,
PRIMARY KEY (id) );
```

In both cases shown here, you specify the keyspace first (USE cycling) to set the current keyspace. You can also use the following notation to identify the keyspace instead of executing the USE statement first:

```
cqlsh> CREATE TABLE cycling.cyclist_name ( id UUID, lastname text,
firstname text, PRIMARY KEY (id) );
```

Defining a Composite Partition Key

Cassandra stores entire rows of data on a node by the partition key. You can distribute data over multiple partitions (multiple nodes) by using a composite partition key. If a column has a lot of data, then the data is too large to store in a single partition.

You can specify multiple columns for the partition key to segment the data into multiple buckets. A composite partition key consists of two or more columns. Instead of grouping all the data into a single partition, you group the data into a set of smaller partitions.

Note   You specify the entire primary key when searching for a specific row and just the partition key when selecting multiple rows from the same partition.

A composite partition key is helpful when you encounter slowdowns in writing data to a node due to heavy writing to a partition. For example, if your app is writing lots of data, you can break up the data into chunks by year/month/day/hour using a four-column partition key.

If you're not using secondary indexes, you must provide all columns of a composite key when querying the table.

The following example shows how to specify two columns of a table in the primary key as a composite partition key:

```
cqlsh> CREATE TABLE cycling.rank_by_year_and_name (
        race_year int,
        race_name text,
        cyclist_name text,
        rank int,
        PRIMARY KEY ((race_year, race_name), rank)
      );
```

The primary key here consists of both a partition key and a clustering column. The partition key is composite, as you use double parentheses around the two columns of the partition key (race_year, race_name). The primary key also has a clustering column, rank.

You must specify all columns of the partition key to retrieve data from this table because you haven't created a secondary index on the table. To get the ranks of cyclists that completed in various races, you must supply values for both the year and race name, as shown here:

```
cqlsh:cycling> SELECT * FROM cycling.rank_by_year_and_name WHERE race_
year=2015 AND race_name='Tour of Japan - Stage 4 - Minami > Shinshu';
```

```
 race_year | race_name                                  |rank|   cyclist_name
-----------+--------------------------------------------+----+-----------------
      2015 | Tour of Japan - Stage 4 - Minami > Shinshu | 1  | Benjamin PRADES
      2015 | Tour of Japan - Stage 4 - Minami > Shinshu | 2  |     Adam PHELAN
      2015 | Tour of Japan - Stage 4 - Minami > Shinshu | 3  |    Thomas LEBAS
(3 rows)
cqlsh:cycling>
```

If you specify only part of the partition key, as shown in the following example, Cassandra complains:

```
cqlsh:cycling> SELECT * FROM cycling.rank_by_year_and_name WHERE race_
year=2016;
InvalidRequest: Error from server: code=2200 [Invalid query]
message="Cannot execute this query as it might involve data filtering and
thus may have unpredictable performance. If you want to execute this query
despite the performance unpredictability, use ALLOW FILTERING"
```

The error is because you used only part of the partition key (race_year) and have omitted the race_name column, which is the second column in the partition key.

You can get pass this error by doing what Cassandra is asking you, by adding the clause ALLOW FILTERING to your query that is using only part of the partition key.

```
cqlsh:cycling> SELECT * FROM cycling.rank_by_year_and_name WHERE race_
year=2015 ALLOW FILTERING;
 race_year | race_name                                  | rank | cyclist_name
-----------+--------------------------------------------+---+------------
      2015 |   Giro d'Italia - Stage 11 - Forli > Imola | 1 | Ilnur ZAKARIN
      2015 |   Giro d'Italia - Stage 11 - Forli > Imola | 2 | Carlos BETANCUR
...
 (5 rows)
```

There may be a performance penalty when you add the ALLOW FILTERING clause to a query. The impact depends on how many rows are in a partition, and if the rows are spread over many SSTable files. However, Cassandra guarantees that the query will be limited to a single node.

Compound Keys and Clustering Columns

When creating a primary key, the clustering column(s) is optional and follows the mandatory partition key of a table. The order of the clustering columns defines the clustering order for a partition in the table. For each partition, Cassandra physically orders the rows based on the clustering order that you specify.

Clustering is how Cassandra sorts data within each partition based on how you define the clustering column(s). By default, Cassandra sorts a column in the ascending order.

A primary key with both a partition key and a clustering column(s) is called a *compound primary key*. The partition key may be simple or composite.

Here's an example that shows how the clustering columns define the clustering order for a partition:

```
CREATE TABLE t (
    a int,
    b int,
    c int,
    PRIMARY KEY (a, b)
);
```

Insert some rows and run the following SELECT statement:

```
SELECT * FROM t;
   a | b | c
 ---+---+---
   0 | 0 | 4      // row 1
   0 | 1 | 9      // row 2
   0 | 2 | 2      // row 3
   0 | 3 | 3      // row 4
```

In this case, the primary key definition specifies *(a, b)*, meaning that column a is the partition key and the column b is the clustering column. In addition, you can see that Cassandra internally stores the rows belonging to the same partition (column a=0) in the order of the values of the clustering column, b.

Clustering a column in the way shown here means that queries that seek a range of rows from a partition can return very quickly. An example is a query such as SELECT * FROM t where a=0 and b>1 and b<=3.

Note   When you specify a compound primary key, Cassandra stores an entire row on a node by its partition key and orders the data with the clustering column(s).

If your application is retrieving data from a large partition, it must normally read the entire partition to get at a small amount of data. If you sort the partition key's data by specifying a clustering column, the database can retrieve data far more efficiently since it won't be reading the entire partition.

Grouping data by specifying a clustering column is similar to how you join tables in a relational database, only better.

In the following table, the column `category` acts as the partition key and the column `points` as the clustering column. When you query this table, for each category the database orders the `points` column in the descending order.

```
CREATE TABLE cycling.cyclist_category (
category text, points int, id UUID, lastname text,
PRIMARY KEY (category, points))
WITH CLUSTERING
ORDER BY (points DESC);
```

Static Columns

You can declare a non-clustering column as *static* when creating a table. A static column is static within a specific partition only.

The following is an example that shows how to declare a static column:

```
CREATE TABLE t (
  k text,
  s text STATIC,
  i int,
  PRIMARY KEY (k, i)
);
INSERT INTO t (k, s, i) VALUES ('k', 'I''m shared', 0);
INSERT INTO t (k, s, i) VALUES ('k', 'I''m still shared', 1);
SELECT * FROM t;
Output is:
 k |                   s | i
----------------------------
k  | "I'm still shared" | 0
k  | "I'm still shared" | 1
```

Things to remember about a static column:

- You can't declare a column acting as the primary key as static.

- A table that doesn't have any clustering columns can't have a static column.

Table Options

You can set two key table options, COMPACT STORAGE and CLUSTERING ORDER, by using the WITH keyword.

Compact Tables

You can define a compact table with the COMPACT STORAGE option. This option exists only for backward compatibility purposes, for dealing with table definitions created before CQL Version 3. You must not specify this option for new tables, so I won't discuss this option in detail.

Clustering Order

The CLUSTERING ORDER option lets you change the clustering order to use the reverse natural order of the columns. By default, the ordering implies ASC (ascendant order). You can specify either the ASC (the default order) or the DESC (descendent order) order.

Other Options

In addition to the two options I described here, a table supports other options such as those that allow you to write comments, specify read repair options, and the time to wait for garbage collecting *tombstones*, which are deletion markers. Here's a brief description of the table options:

- comment: A comment about the table.

- read_repair_chance: The probability with which to query more nodes than those required by the consistency level for performing read repairs. Default value is 0.1

- dclocal_read_repair_chance: The probability with which to query more nodes than those required by the consistency level belonging to the same data center for performing read repairs. Default value is 0.

211

- gc_grace_seconds: Time to wait before the garbage collection of the tombstones. Default value is 864000 seconds.

- bloom_filter_fp_chance: The target probability of the false positives of the SSTable Bloom filters.

- default_expiration_time: The time-to-live (TTL) for a table in seconds.

There are three very important performance-influencing table options that relate to compaction, compression, and caching. I'll explain all three of them in detail in Chapter 11, which discusses Cassandra performance tuning. You can read ahead if you wish.

Using a Counter Column to Track Values

You can use a special column called a *counter* to keep a tally of a value that's incremented. For example, you may want to create a counter column to trace the number of online game players who joined a game or the number of page views or log messages.

You must specify the counter data type for a counter column. However, you can't denote a column in any table as a counter column; you must create a dedicated table to hold a counter column. The dedicated table must include only a primary key (could be composite) and the counter column. You can't use the counter column as a primary key; all columns other than the primary key column must be counter columns.

Note   You can't index a counter column. You also can't delete it.

You can't use the INSERT statement to load data into a counter column; instead, you must use the UPDATE statement to both insert and modify a counter column's values.

The following example shows how to create a counter column:

```
CREATE TABLE popular_count (
    id UUID PRIMARY KEY,
    popularity counter
    );
```

Altering a Table

You can modify a table structure or its options by executing an ALTER TABLE statement. The ALTER TABLE statement allows you to do the following.

- You can change the type of a column, as shown here:

```
ALTER TABLE customers ALTER lastKnownAddress TYPE UUID;
```

- You can add a new column to the table.

```
ALTER TABLE customers ADD address varchar;
```

- You can also change some of the table options, by adding the WITH instruction.

```
ALTER TABLE customers
      WITH comment = "complete customer records table'
      AND read_repair_chance = 0.4;
```

Note   You can't change the COMPACT STORAGE and CLUSTERING ORDER options once you create a table.

You can convert some CQL data types only as allowed. Please refer to the CQL documentation for the CQL type compatibility. For example, you can convert an existing ASCII or text type to only the varchar type.

Viewing a Table's Configured Options

As I showed earlier, you can specify many options when creating a table. You can also modify several options by altering a table. How do you know which options a table has at any given time? Run the describe table command to view all properties of a table.

In the following example, you create the table cyclist_name and then run the command describe cyclist_name to view the table's properties:

```
cqlsh> use cycling;
cqlsh:cycling> CREATE TABLE cyclist_name ( id UUID, lastname text,
firstname text, PRIMARY KEY (id) );
cqlsh:cycling> describe cyclist_name;
```

```
CREATE TABLE cycling.cyclist_name (
    id uuid PRIMARY KEY,
    firstname text,
    lastname text
) WITH bloom_filter_fp_chance = 0.01
    AND caching = {'keys': 'ALL', 'rows_per_partition': 'NONE'}
    AND comment = ''
    AND compaction = {'class': 'org.apache.cassandra.db.compaction.
    SizeTieredCompactionStrategy', 'max_threshold': '32', 'min_threshold': '4'}
    AND compression = {'chunk_length_in_kb': '64', 'class': 'org.apache.
    cassandra.io.compress.LZ4Compressor'}
    AND crc_check_chance = 1.0
    AND dclocal_read_repair_chance = 0.1
    AND default_time_to_live = 0
    AND gc_grace_seconds = 864000
    AND max_index_interval = 2048
    AND memtable_flush_period_in_ms = 0
    AND min_index_interval = 128
    AND read_repair_chance = 0.0
    AND speculative_retry = '99PERCENTILE';
```

```
cqlsh:cycling>
```

Of the many properties of this table, you've set only a few. The listing is long, however, since the dataset uses *default values* for all the table options that you didn't specify.

Dropping and Truncating Tables

As with any relational database, you can drop and truncate Cassandra tables.

Dropping a Table

You can drop a table with the DROP TABLE statement.

```
cqlsh> DROP TABLE cycling.cyclist_name;
cqlsh> DROP TABLE cycling.cyclist_name IF EXISTS;
```

Before you drop a table, you must first drop any materialized views that are based on the table.

Truncating a Table

Truncating a table with the TRUNCATE statement leaves the table in place but removes all the data in the table, including the data from any materialized view based on the table.

```
cqlsh> TRUNCATE TABLE customers;
cqlsh> TRUNCATE customers;
```

Both of these commands work the same way. The keyword TABLE is optional since a table is the only object you can truncate.

Before you truncate a table, you need to do the following:

1. Set the consistency level to ALL with the CONSISTENCY command.

    ```
    cqlsh> CONSISTENCY ALL;
    Consistency level set to ALL.
    cqlsh>
    ```

2. Run the nodetool status command to ensure that all nodes are up and available.

    ```
    cqlsh> nodetool status
    Datacenter: datacenter1
    =======================
    Status=Up/Down
    ```

```
|/ State=Normal/Leaving/Joining/Moving

--  Address    Load        Tokens  Owns (effective)  Host ID              Rack
UN  127.0.0.1  159.91 KiB  256     100.0%            c51011d6-06da-47b8-bd55-
                                                     70ac2e51716d  rack1

$
```

Deleting Rows and Columns from a Table

You can remove rows from a table with the DELETE statement.

cqlsh:mydb> DELETE FROM cyclists where pk = 'salapati';

You can remove rows from a table by issuing the DELETE statement in the following way:

cqlsh:mydb> DELETE session-token FROM users where pk = 'salapati';

This statement will remove the column session_token from just one row where the value of pk is 'salapati'. The WHERE clause specifies which row/rows to delete from a table.

Deleting Multiple Rows

You can delete multiple rows by specifying the keyword IN and supplying a comma separated list of values in parentheses.

```
cqlsh> DELETE FROM cycling.cyclist_name WHERE firstname IN ('Sam', 'James');
```

If you wish to remove an entire column from a table, you can do so by leaving out the WHERE clause.

```
cqlsh:mydb> DELETE session_token FROM users;
```

Deleting from a Collection Set, List, or Map

To remove an element from a map that you stored as a single column, you need to specify the column's name along with the element's key in square brackets.

```
cqlsh> DELETE sponsorship ['sponser_name'] FROM cycling.races WHERE
race_name = 'Criterium du Dauphine';
```

To delete data from a list, specify the name of the column along with the list index positon.

```
cqlsh> DELETE categories[3] FROM cycling.cycling.history WHERE lastname =
'TIRALONGO';
```

To remove all elements from a set, just specify the name of the column.

```
cqlsh> DELETE sponsorship FROM cycling.races WHERE race_name = 'Criterium
De Dauphine';
```

Deleting Old Data by Using a TIMESTAMP

You can specify columns for deletion using a timestamp to identify the column.

```
cqlsh> DELETE  firstname, lastname
       FROM cycling.cyclist_name
       USING TIMESTAMP 1318452291034
       WHERE lastname = 'ALAPATI';
```

Removing the Deleted Data from the Database

Cassandra doesn't immediately remove the rows and columns from the table once you issue your DELETE statements. Rather, it removes the values completely during the first compaction following your delete operation. This is to improve performance.

How Cassandra Uses Tombstones to Mark Deleted Data

Cassandra marks the data you delete with a tombstone and removes it completely after a grace period. Cassandra treats all deletes as either an insert or an upsert (update or insert, depending on the case).

When you delete data, Cassandra adds a deletion marker to the partition that's part of the DELETE command. This deletion marker is called a tombstone, and the database writes these markers to an SSTable on one or more nodes. Each tombstone has an expiration period, after which the database deletes the tombstone as part of its routine SSTable compaction process.

Specifying a Time-to-Live Value

Alternatively, you can specify a TTL value for a record (row or a column) or a table to set an optional expiration period for data in a column (other than counter data). Doing so will make Cassandra delete the data in the column or the table after a certain period has elapsed. At the end of the time-to-live period, the database marks the record with a tombstone and takes care of it as explained earlier, during a compaction process.

Expiring data by setting a TTL doesn't come free of cost. It involves an additional overhead of 8 bytes in memory and on disk to record the TTL and expiration time.

Note   Specifying a TTL is a way of implicitly deleting data. You can do this by either setting the TTL on columns in a table or by specifying the default_time_to_live property for a table.

You can specify the expiration time (TTL) with an INSERT or an UPDATE statement. The following example shows how to specify the password column in the users table to expire after 1 day (86,400 seconds) by adding the USING TTL option to the INSERT command):

```
cqlsh:mydb> INSERT INTO users
            user_name, password)
            VALUES ('cbrown', 'ch@ngem4a') USING ttl 86400;
```

Following this, you can increase the period to 2 days by running the following UPDATE statement:

```
cqlsh:mydb> UPDATE users USING TTL 172800 SET password = 'ch@ngem4a'
            WHERE user_name = 'cbrown';
```

You can set a default TTL on an entire table with the default_time_to_live property, which is one of the CQL table properties you can specify when creating a table. The default value of this property is 0, meaning that data you write to a table will never expire. You set the value in number of seconds. Once you set it, Cassandra applies a TTL value to each column in a table. Cassandra will then tombstone the entire table once the table's TTL is exceeded.

Note   By setting the default_time_to_live table property to zero, you are in effect removing any column TTLs that you've specified in that table.

You can apply the default TTL to a table when creating a table with the CREATE TABLE statement or when modifying it later with an ALTER TABLE statement.

Once the number of seconds since the creation of a column crosses the TTL value you've configured, Cassandra considers the data as expired, although it reports it in the query results. Cassandra marks the expired data with a tombstone, which is a deletion marker, on the next read but retains the data for a maximum of gc_grace_seconds.

Once the time specified by the gc_grace_seconds property passes, Cassandra automatically removes the data that it has marked with a tombstone during its regular compaction processes.

How Garbage Collection Works

You set the gc_grace_seconds property when creating a table. This property specifies the number of seconds after Cassandra marks data with a tombstone before the data becomes eligible for garbage collection.

The default value for the gc_grace_seconds property is long (864,000 seconds, which is 10 days), so you can leave this property alone when creating your tables. The reason the default value is so long is to enable Cassandra to have plenty of time to maximize consistency of data before permanently getting rid of it.

Although the default value of the gc_grace_seconds property is 10 days, you can reduce it for tables whose data your users won't be explicitly deleting, such as tables that contain just data with their TTL configured, or a table for which you've configured the default_time_to_live property.

Tip   You can configure each table with a different grace period for a tombstone.

Before you lower the value of the gc_grace_seconds property from its default setting of 10 days, remember the following:

- Cassandra doesn't replay *hints*, which are instructions to replay missed write operations, that are older than the gc_grace_seconds setting. This means that reducing the gc_grace_seconds property value may result in a recently revived node missing some writes.

- Cassandra also performs *batch operations,* which involve sequentially replaying changes in the database. Cassandra will wait to replay a batched change until gc_grace_seconds elapse after the batched change was created. If you decrease the value of the gc_grace_seconds property, you run the risk that a batched write will restore some deleted data.

Zombie Records and the Need for Node Repair

When you delete data, the node that processes the delete request will create a tombstone for the record and pass the tombstone to the other nodes that are storing a replica of the deleted record. If one of the replicas happens to be down, it won't be able to receive the tombstone and will continue to store the pre-deleted version of the data.

The database considers the (deleted) record on the node that has recovered as fresh data, if the database deleted the tombstoned record while the node was down. The database will then send this "fresh" data around to the other nodes. This deleted row(s) that reappears in a table after deletion is called a *zombie*.

The reason for the presence of zombie data is that a node came back up after being down for a long period, before the database could run a repair. If the node isn't repaired before it is back online, the database will see the non-tombstoned data and copy it to the other nodes as new data. This is the reason you must run a repair on a node that you've restored (with the `nodetool repair` command) before allowing it to join the cluster.

Note   The database doesn't replay a mutation (insert or update) for a tombstoned record that is within its grace period.

The table property `gc_grace_seconds` sets the grace period for a tombstone. This grace period helps keep the zombie records from popping up, by letting a recovering node time to process tombstones in the normal fashion. When you recover the down node, the database replays all the mutations (inserts and updates) that this node missed while it was down, via its hinted handoffs feature. The database won't replay the changes for a record marked for deletion with a tombstone until its grace period is over. If the node fails to come back up by the end of the grace period, it'll miss the deletions.

Once the grace period for a tombstone ends, the database deletes the tombstone during its normal compaction operations.

You can prevent zombie records by

- Running the `nodetool repair` command on the node once it recovers

- Running the `nodetool repair` command on each table every `gc_grace_seconds`

Note   A DROP KEYSPACE or a DROP TABLE command will immediately delete data.

Note the following:

- If you choose to specify a TTL for all records in a table, they'll expire naturally and you don't need to run the nodetool repair command for that table.

- You can immediately delete all tombstones by initiating a compaction process (when you use the SizeTieredCompactionStrategy to compact data).

- When you set the default_time_to_live property at the table level, when a record exceeds the table's TTL, Cassandra removes the record without creating a tombstone or waiting for the compaction process to run.

Indexing in a Cassandra Database

A primary index is mandatory for a table, and you've already learned how to create and work with a primary index.

A *secondary index* is not mandatory when creating a table. A secondary index enables you to query a table on a column that you won't be able to query otherwise.

A secondary index helps filter a table's data in a non-primary key column. Cassandra doesn't allow you to run a query that matches a non-primary key column because it may not result in retrieving a continuous chunk of data from the table.

Cassandra spreads a table's data across multiple partitions stored on various nodes since a non-primary key isn't involved in ordering the data. This means that if you query a specific value of a non-primary key column, you many end up reading all partitions of table; therefore Cassandra doesn't permit this type of access.

You can build a secondary index on a column, but there are some limitations as to how you can make use of such an index. If your query includes a partition key in addition to a secondary index column, it'll work since the database can satisfy this query by reading data from a single node partition. However, if you don't restrict the secondary index in a query to a specific partition key, the query will need to read data from all the nodes and the database won't allow it. You'll see the following error when you try the query:

```
InvalidRequest: Error from server: code=2200 [Invalid query]
message="Cannot execute this query as it might involve data filtering and
thus may have unpredictable performance. If you want to execute this query
despite the performance unpredictability, use ALLOW FILTERING"
```

For this type of query to work, you must add the option ALLOW FILTERING to the query.

When to Use an Index

Use an index when a table has many rows with the indexed value. If a table has too many unique values for a column, you incur more of an overhead for querying and maintaining the index on that column.

Let's say you have a table named races, and there are hundreds of millions of entries for cyclists that participate in various bicycling races over several years. Your goal is to find the cyclists by rank. Since many of the cyclists have participated during the same years, their ranks will share the same column value for the race year. In this case, the race_year column in the table races is ideal as an index on that table.

Note   The SSTable Attached Secondary Indexes (SASO) index are a new implementation of secondary index that offer a performance improvement. However, SASI indexes are still experimental as of the writing of this book (October 2017).

When Not to Use a Secondary Index

An index isn't always a boon. It can easily turn out to be a curse if you indiscriminately create indexes on all tables. You're advised not to create indexes when dealing with

- High and low cardinality columns

- Columns that are frequently updated or deleted

- Tables that include a counter column

- When searching for a row in a large partition

In the following sections, I elaborate the reasons for the inadvisability of indexes in the situations listed here.

High and Low Cardinality Columns

Neither a high cardinality column nor a very low cardinality column is a suitable candidate for an index. High cardinality means that the column has many distinct values, and low cardinality means there are few distinct values.

It's inefficient to create and use an index for a column with numerous distinct values. You're better off maintaining the table as a form of an index rather than creating a built-in Cassandra index.

If your table is large but the queries are rare on the table, you may opt to create an index on a high cardinality column. Just don't do it on a highly used, large table.

As for very low cardinality columns, take the extreme case of a column with Boolean values. In this case, each value of the index is a single row in the index, and the rows are extremely large since there are only two possible values for each indexed column.

Frequently Updated or Deleted Columns

As mentioned, Cassandra uses tombstones to mark deleted data. Cassandra stores the tombstone in the index until it hits a limit of 10,000 cells. Once Cassandra exceeds the tombstone limit, any query that refers to the indexed column will fail.

Using an Index for Searching for Rows in Large Partitions

If you query an indexed column in a large table, it usually requires gathering the responses from several partitions. As you add more nodes to your cluster, the query response gets slower and slower.

The way to avoid the situation described here is to narrow the search when querying the indexed column.

In addition to all the potential issues described here, since the database stores the index table on each node of the cluster, if a query accesses multiple nodes, a query can turn out to be a performance problem.

Creating a Secondary Index

You can create a secondary index on a table by executing the CREATE INDEX statement. Let's say you have the following table, with a primary key that consists of a composite partition key (race_year, race_name) and a clustering column (rank):

```
cqlsh> CREATE TABLE cycling.rank_by_year_and_name (
        race_year int,
        race_name text,
        cyclist_name text,
        rank int,
        PRIMARY KEY ((race_year, race_name), rank)
        );
```

You can't query this table on the column race_year since that column is only a part of the composite partition key. You can create a secondary index on the race_year column, which enables you to query the table on this column. Here's how you create the index:

```
cqlsh> CREATE INDEX ON cycling.rank_by_year_and_name (race_year);
```

You can now issue the following query that makes use of your new secondary index:

```
cqlsh SELECT * FROM cycling.rank_by_year_and_name WHERE race_year=2015;
```

In the CREATE INDEX statement, you didn't specify a name for your new index. Naming an index is optional, and if you don't specify one, the database will assign a system generated name for the index. You can alternatively specify your own index name thus:

```
cqlsh> CREATE INDEX my_index1 ON cycling.rank_by_year_and_name (race_year);
```

You can specify any of the table's columns for a secondary index, including a clustering column. You can create multiple secondary indexes on a table.

You can create an index on a table that already contains data, as well as a new table with no data.

Dropping an Index

Unlike in the case of a primary index, you can drop a secondary index. Issue the DROP INDEX statement to drop an index:

```
cqlsh> DROP INDEX myIndex1;
```

This statement drops an existing secondary index. If you wish to avoid errors since the index doesn't exist, you can use the following statement instead:

```
cqlsh> DROP INDEX myIndex1 IF EXISTS;
```

Materialized Views in Cassandra

As mentioned, secondary indexes aren't good when dealing with high cardinality data since queries on these indexes make the database access all the nodes in the cluster. Materialized views are an appropriate alternative for handling high cardinality data.

A materialized view is a table that you build from an existing table using a new primary key. You can include rows with null values in the primary key column.

Tip   Materialized views hinder performance when you use them for querying low cardinality data. However, please proceed with caution because materialized views still require work to be fully useful. It is not unheard of for MVs to cause cluster downtime when dealing with heavy updates.

MVs should be much more usable by the time Cassandra 3.11 or 4.x rolls around.

Creating a Materialized View

The following is an example that shows how you create a materialized view using a regular table as its basis or source:

```
cqlsh> CREATE MATERIALIZED VIEW cycling.cyclist_by_age
       AS SELECT age, name, country
       FROM cycling.cyclist
       WHERE age IS NOT NULL AND cid IS NOT NULL
       PRIMARY KEY (age, cid);
```

In this example,

- The AS SELECT clause specifies the columns that you want to copy from the base or target table to the new materialized view. Here they are *age, name*, and *country*.

- The FROM clause points to the original or source table from which the materialized view draws its data.

- The WHERE clause ensures that the primary key columns are non-null since that's a requirement for creating a materialized view.

- The PRIMARY KEY clause specifies the primary key columns, which are age and cid in this example. The column age also happens to be the primary key column for the source table, and you must include all such primary key columns from the source table in the primary key of the materialized view as well.

Dropping a Materialized View

You can drop a materialized view by issuing the DROP MATERIALIZED VIEW command:

```
cql> DROP MATERIALIZED VIEW cycling.cyclist_by_age;
```

You can alternatively drop a materialized view or views by truncating the source table, which removes all materialized views based on that table. When you drop a table, you must first drop all materialized views that use that table as their source table.

Denormalizing with Materialized Views

In the example of a materialized view, you partitioned the materialized view by age so you could run queries on it based on the age of the cyclists. Materialized views help denormalize your data by creating multiple tables with the same data but organized differently based on different primary keys.

You can create multiple materialized views on the same table. Thus, you can create view with different primary keys such as birthday or country. This helps you organize the data by these primary keys and lets you query those materialized views by *birthday* and *country*.

Updating Materialized Views

When you update the source table, the database automatically updates any materialized views you've created based on that table, and the same goes with deletion of data from the source table. However, the updates are asynchronous so the materialized view is updated after a delay, after the update of the parent table.

Cassandra needs to perform an additional read before it can update the materialized views. This read involves a data consistency check of all the replicas, which means that writes to a materialized view are slower than normal table writes.

Deletes pose a problem as well. Since the database may not store identical rows from the parent table contiguously in the materialized views, deleting rows from the source table may often require the database to create multiple tombstones in the corresponding materialized views.

Data manipulation language (DML) is the set of CQL statements that enable you to query, insert, update, and delete data. In the following sections, I'll review the essentials of the Cassandra DML statements.

Querying Data with the SELECT Statement

Querying data is the most common DML operation that you'll perform in a Cassandra database. Therefore, it's a good idea to thoroughly understand how to use the SELECT statement in CQL.

The SELECT statement returns results that match a request. Each of the rows contains the values for the selections that are part of the SELECT statement.

The following is the basic syntax of a SELECT statement in CQL:

```
SELECT selectors | DISTINCT partition
FROM [keyspace_name.] table_name
[WHERE partition_conditions [ AND solr_query = 'search_expression' [LIMIT n] |
   [solr_query = 'search_expression'  [LIMIT n]]
   [AND clustering_conditions
   [AND regular_column_conditions]]]
[GROUP BY column_name]
[ORDER BY PK_column_name ASC|DESC]
[LIMIT N | PER PARTITION LIMIT N]
[ALLOW FILTERING
```

A SELECT statement contains the following key clauses:

- The *selection* clause

- The *where* clause

- The *group by* clause

- The *ordering* clause

The following sections describe the key clauses of the SELECT statement.

The Selection Clause

The selection clause determines which columns the query returns, as well as any transformation the database must apply to the result set before returning the results. The SELECT statement must use either a wildcard selector (*) or one or more selectors to define the columns.

A selector can be

- A column name

- A term

- A function call

- The special call COUNT(*) to the COUNT function

Here are some typical SELECT statements:

```
cql> SELECT COUNT (*) FROM users;
cql> SELECT name, occupation FROM users;
cql> SELECT name, occupation FROM users where userid in (1111,1112,111130);
cql> SELECT JSON name, occupation FROM users WHERE userid = 999;
```

You can alias a top-level selector by specifying the AS clause in a SELECT statement, as shown here:

```
cql> SELECT name AS customer_name FROM customers;
Cutomer_name
Alapati
....
```

The WHERE Clause

The WHERE clause in a CQL SELECT statement has some surprises for those of us that come from a relational database background. As is the case with the other databases, the WHERE clause specifies the rows that the database must query. However, the following rules apply:

- The WHERE cause can only consist of relations on columns that are part of a PRIMARY key or you must define a SECONDARY INDEX on the columns.

- For any partition key, the relations on the clustering columns are restricted to just those relations that seek a contiguous set of rows.

Let's say you've created the following table and then inserted some data into it:

```
cqlsh> create table users ( id UUID PRIMARY KEY, lastname text, firstname
text );
cqlsh> insert into users (id, firstname, lastname)
          ... values
          ... (uuid(), 'sam', 'alapati');
```

The following query will work for sure since you specify the primary key, the id column:

```
cqlsh> select * from users where id=5361a682-9aea-46df-91b0-40572cfa9c97;
 id                                   | firstname | lastname
--------------------------------------+-----------+----------
 5361a682-9aea-46df-91b0-40572cfa9c97 |       sam |  alapati
(1 rows)
cqlsh>
```

However, the following query won't work:

```
cqlsh > select * from users where lastname='alapati';
InvalidRequest: Error from server: code=2200 [Invalid query]
message="Cannot execute this query as it might involve data filtering and
thus may have unpredictable performance. If you want to execute this query
despite the performance unpredictability, use ALLOW FILTERING"
```

The reason this query won't work is that there's no secondary index on the `lastname` column, and the column doesn't select a contiguous set of rows. You need a secondary index on the `lastname` column if you wish to query on that column.

Creating a Secondary Index

You can make this query work by creating a secondary index on the `lastname` column. A secondary index is an index on any column in a table other than the keys that are part of the primary key.

Once you create the secondary index on the column `lastname`, the previous query works fine:

```
cqlsh:cycling> create index on users(lastname);
cqlsh:cycling> select * from users where lastname='alapati';
 id                                   | firstname | lastname
--------------------------------------+-----------+----------
 5361a682-9aea-46df-91b0-40572cfa9c97 |       sam |  alapati

(1 rows)
Cqlsh:cycling>
```

In addition to creating secondary indexes on simple columns, such as shown here, you can also create indexes on collection-based columns such as the map-, set-, and list-based collections.

Drawbacks of Secondary Indexes

Secondary indexes are somewhat of a mixed blessing due to several reasons. For example, all nodes must maintain a local copy of the secondary index based on the data that's in the partitions on that node. Since the query on a secondary index often needs to traverse multiple nodes, it's not cheap to perform the query.

Secondary indexes are not suitable for columns with too high or too low cardinality. Cardinality is the number of distinctive values for a column. The indexes are also not good for handling data that you frequently modify or delete, the reason being that the database must deal with all the tombstones generated by the updates and deletes before the compaction process can take care of all them.

Ideally, you should create multiple tables to cater to the different queries that you want to execute on a chunk of data. Unlike in the case of relational databases where secondary indexes are an essential part of querying, Cassandra's secondary indexes are at best a backup strategy, and you're better off creating materialized views, which follow the recommended pattern of denormalization. I discuss materialized views in the "Working with Materialized Views" section.

SASI: New Implementation of Secondary Indexes

Cassandra 3.4 introduced a newer implementation for secondary indexes called the SSTable Attached Secondary Index. Unlike the traditional secondary indexes which are stored in "hidden" tables, the database stores the SSI indexes as part of an SSTable file.

You can create an SASI index with the CREATE CUSTOM INDEX command:

```
cqlsh> CREATE CUSTOM INDEX on users (lastname)
       USING 'org.apache.cassandra.index.sasi.SASIIndex';
cqlsh>
```

You can use both traditional and SASI indexes together. The newer secondary indexes allow you perform inequality searches and use the LIKE keyword to perform text searches, both of which you can't do with traditional indexes.

You can group CLUSTERING COLUMNS together in a relation by employing the *tuple* notation, as I explain later in this chapter.

Writing a Conditional Statement

A query can scan a table's partitions to retrieve a segment of data. For this type of query to work, you must store the segment sequentially so that you can use the clustering columns to define the segment of data you need.

Let's use a simple example to show how to write a conditional statement. The table race_times shows the race times of cyclists for various races:

```
cqlsh> CREATE TABLE cycling.race_times (race_name test, cyclist_name text,
       race_time text,
       PRIMARY KEY (race_name, race_time);
```

Here `race_name` is the primary key and `race_time` is the clustering column. You can now specify a conditional operator to find a segment of data. You can do this using the `race_time` column, as shown here:

```
cqlsh> SELECT * FROM cycling.race_times
        WHERE RACE_NAME = '17th Santos Tour Down Under'
        AND race_time >='19:15:19"'
        AND race_time <= '19;15:39');
```

How to Group a Query's Results with the GROUP BY Clause

Specify the GROUP_BY clause to gather into a single row all rows that share the same values for a column (or a set of columns).

When you specify an aggregate function such as AVG along with a GROUP_BY clause, Cassandra generates a separate value for each group. Otherwise, specifying an aggregate function results in a single value for all rows.

Here's an example showing how to specify the GROUP_BY clause:

```
cqlsh> SELECT weatherstation_id, date, MAX(temperature)
        FROM temperature_by_day
        GROUP BY weatherstation_id, date;
```

In this example, the table's partition key includes two columns, `weatherstation_id` and `date`, and the clustering key is `event_time`. The GROUP BY uses the partition key (`weatherstation_id`, `date`).

If I want to, I can use the GROUP BY with both the partition key and the clustering key, as shown here:

```
cqlsh> SELECT weatherstation_id, date, MAX(temperature)
        FROM temperature_by_day
        GROUP BY weatherstation_id, date, event_time;
```

Ordering a Query's Results with the ORDER BY Clause

You can fine-tune the order in which Cassandra displays the results of a SELECT statement. Specify the ORDER BY clause to order the results of a query. You can only choose the columns that you define on the table with the clustering order column.

You must define the partition key in the WHERE clause and define the clustering column for ordering results by specifying the ORDER BY clause.

The following is an example that shows how to use the ORDER BY clause.

First, create the following table:

```
cqlsh>  CREATE TABLE cycling.cyclist_cat_pts
            (CATEGORY text, points int, id UUID, lastname text,
            PRIMARY KEY (category, points) );
```

Now run the SELECT statement with the ORDER BY clause.

```
cqlsh> SELECT * FROM cycling.cyclist_cat_pts WHERE category = 'GC'
        ORDER BY points;
```

Using the ORDER_BY clause in cqlsh requires you to turn off paging with the PAGING OFF command.

```
cqlsh> PAGING OFF
```

Limiting Your Results

You can limit the number of rows output by a SELECT statement by specifying the LIMIT *N* option with a SELECT statement, as shown here:

```
cqlsh> SELECT * FROM cycling.cyclist_name LIMIT 3;
```

You can also specify a PER PARTITION LIMIT *N* option to limit the number of rows returned for a specific partition. First, create a table that'll sort the data into multiple partitions and insert data into that table. In the example here, you use the race_year and race_name columns as the composite partition key and rank as the clustering column:

```
cqlsh> CREATE TABLE cycling.rank_by_year_and_name (
        race_year int
        race_name text.
        cyclist_name text,
        rank int,
        PRIMARY key ((race_year, race_name), rank);
```

With this table in place, you can execute the following SELECT statement. The PER PARTITION LIMIT N clause retrieves the top five races for each combination of the two columns in the partition key, race_year and race_name.

```
cqlsh> SELECT * FROM cycling.rank_by_year_and_name PER PARTITION LIMIT 5;
```

Filtering the Results

Cassandra doesn't allow queries that require filtering. It only allows queries that don't require filtering, where it returns all the records read in the result set. The reason for this is so queries have predictable performance. A query will always execute in a time that's proportional to the amount of data the query returns.

Let's say you have the following table:

```
CREATE TABLE emails (
    emailId int,
    time int,
    from text,
    content text,
    PRIMARY KEY(emailId, time));
```

You can retrieve all data from the table by issuing the following query:

```
SELECT * from emails;
```

However, if you issue the following query, Cassandra complains:

```
SELECT * FROM emails WHERE time1 = 1418306451235;
Bad Request: Cannot execute this query as it might involve data filtering
and thus may have unpredictable performance. If you want to execute this
query despite the performance unpredictability, use ALLOW FILTERING.
```

The reason for the error is that Cassandra doesn't run a query when it's unable to guarantee that it won't be scanning a large amount of data even when there are only a handful of values for the querying. Cassandra is looking out for you, trying to conserve your computing resources.

To execute the query, the database needs to retrieve all rows from the table EMAILs and then filter out the rows that don't have the requested value for the TIME column. However, if you know that over 90% of the rows in the table have the requested value for the time column, your query is going to efficient, and you must specify the ALLOW FILTERING clause in your query.

If the table contains a million rows and only five rows contain the requested value for the time column, your query will waste resources, since Cassandra will be reading all million rows to retrieve five rows. In this case, you should think of adding a secondary index. So, you can add an index on the FROM column and run the following query:

```
SELECT * FROM emails WHERE from = 'Sam Alapati';
```

Cassandra will retrieve all emails sent by Sam Alapati and won't ask you to use the ALLOW FILTERING clause, since it uses the secondary index on the from column to find the matching rows, without having to filter the results.

You know your data better. You can override the default behavior by specifying the ALLOW FILTERING clause and make the same query run successfully.

Specifying Built-In Functions for Aggregating Results

Cassandra 3.1 provides built-in standard aggregate functions such as min, max, avg, sum, and count to aggregate results. I provide examples for specifying two aggregate functions: sum and count.

In the table cycling.cyclist_points, the PRIMARY KEY is (id, race_points). You can find the sum of race points for a specific cyclist with the help of the sum function.

```
cqlsh> SELECT sum(race_points)
       FROM cycling.cyclist_points
       WHERE id=e3b19ec4-774a-4d1c-9e5a-decec1e30aac;
```

In the table cycling.country_flag, the PRIMARY KEY is (country, cyclist_name). You can employ the count function to find the number of cyclists from Belgium.

```
cqlsh> SELECT count(cyclist_name) FROM cycling.country_flag WHERE
country='Belgium';
```

You can apply any custom user-defined function as part of a query, as shown here, where fLog is an UDF created to retrieve data from this table:

```
cqlsh> SELECT id, lastname, fLog(race_points) FROM cycling.cyclist_points;
```

Formatting the Query Results as JSON

You can retrieve a table's data in the JSON format. To get the entire result of a query in the JSON format, just insert the keyword json right after the SELECT statement.

```
cqlsh> SELECT json name, checkin_id, timestamp from checkin;
```

If you wish to retrieve only some column(s) in the JSON format, you can do so by enclosing the column's name in toJson().

```
cqlsh> select name, checkin_id, toJson(timestamp) from checkin;
```

Selecting Data from a Collection Column

You query a collection column the same as any other column. The following example shows how you retrieve data for a specific cyclist ID. Here, the column teams is a set:

```
cqlsh> SELECT lastname, teams
       FROM cycling.cyclist_career_teams
       WHERE id = 5b6962dd-3f90-4c93-8f61-eabfa4a803e2;
```

Cassandra will return the complete collection when you query a table with a collection. The results will be in the order based on the element types. For example, text elements will be in alphabetical order. If the key is an integer type, the order will be based on the key values. If you need the database to return the results in the order the data was inserted, use a list instead of a set.

Doing a Multi-Get of CQL Rows with the IN Keyword

You can specify the IN keyword to define a set of clustering columns you want to retrieve together, making it get multiple rows.

As with the ORDER_BY clause, you must turn off paging in cqlsh with the PAGING OFF command when you specify the IN keyword. Here are some same queries that show how to specify the IN keyword.

The following query retrieves and sorts the results based on the clustering column category_id:

```
cqlsh> select * from cycling.cyclist_cat_pts
       WHERE cataegory_id IN (Time-tral', 'Sprint')
       ORDER BY id DESC;
```

The following example shows how to specify multiple clustering columns. Here the race_id column is the partition key, and the race_start_date and race_end_date columns are the clustering columns:

```
cqlsh> SELECT * FROM cycling.calendar WHERE race_id =101
       AND (race_start_date, race_end_date)
       IN (('2015-05-09', '2015-05-31'), ('2015-05-06', '2015-05-31'));
```

Inserting Data with the INSERT Statement

You execute the INSERT statement to write one or more columns for a row in a table. You must specify at least the column on which you've defined the primary key, as otherwise you won't be able to identify the row.

Here's the generic syntax of the INSERT statement:

```
INSERT INTO [keyspace_name.] table_name (column_list)
VALUES (column_values)
[IF NOT EXISTS]
[USING TTL seconds | TIMESTAMP epoch_in_microseconds]
```

Here's an example showing how to perform an insert with the help of the VALUES clause:

```
cqlsh> INSERT INTO cycling.cyclist_name (id, lastname, firstname)
       VALUES (5b6962dd-3f90-4c93-8f61-eabfa4a803e2, 'VOS','Marianne');
```

When you use the JSON syntax, specifying the VALUES clause is optional, but make sure you specify the keyword JSON after the table name.

```
cqlsh> INSERT INTO cycling.cyclist_category JSON '{
       "category" : "GC",
       "points" : 780,
       "id" : "829aa84a-4bba-411f-a4fb-38167a987cda",
       "lastname" : "SUTHERLAND" }';
```

Tip If a row doesn't exist, Cassandra creates the row, and if the row exists already, it updates that row. You really can't tell which one of the two operations—creation or update—took place. You can specify the IF NOT EXISTS condition to insert data only if it doesn't exist, but this has a discernible performance penalty so you shouldn't use it routinely.

The clause UPDATE_PARAMETER in an INSERT statement also occurs in the UPDATE, and DELETE statements, so it's worth understanding this parameter well. The UPDATE_PARAMETER clause supports the following parameters:

- TIMESTAMP: Sets the timestamp for the operation. By default, the coordinator uses the current time at the start of the statement execution as the timestamp.

- TTL: This parameter specifies a time-to-live (in seconds) for the values you insert. This is an optional parameter, and by default values don't expire. Setting the TTL to the value 0 is the same as specifying no TTL.

Modifying Data with the UPDATE Statement

Cassandra's UPDATE capability is really an UPSERT. When you update a row, Cassandra will create the row if didn't exist before, and update it otherwise. The UPDATE statement will thus end up writing one or more column values of a row or rows.

If you insert a row with the same primary key as a current row, the database replaces the row. If you update a row and the primary key doesn't exist, Cassandra creates the row. Because Cassandra employs an append model, both insert and update operations work the same way; there's no essential difference between the two operations.

You can specify the TTL seconds or Timestamp microseconds as options on all columns except counter columns.

You can update a column in multiple rows, as in the following example:

```
cqlsh> UPDATE users
       SET state = 'TX'
       WHERE user_id IN (12345, 23456, 34567);
```

You can update one or more columns at once. The following command shows how to update multiple columns in the same row:

```
cqlsh> UPDATE users
       SET name = 'Sam Alapati',
       email='samalapati@gmail.com'
       WHERE user_id = 23456;
```

You can update data in a collection set, map, or list.

You can also perform conditional updates, as shown here:

```
cqlsh> UPDATE users SET id = 12345
       WHERE lastname = 'ALAPATI' and firstname = 'Sam'
       IF age =50;
```

Conditional updates are lightweight transactions and impose a performance penalty and therefore you must use them judiciously.

CHANGE DATA CAPTURE (CDC) LOGGING

Cassandra offers change data capture (CDC) logging to track data that has changed. You configure CDC logging on a per-table basis.

When you decide to use CDC logging, you must also specify a limit on the amount of disk space the CDC logs can use. When the database flushes the memtable data to disk, the database moves the commit log segments that contain data relating the tables for which you've enabled CDC logging to the CDC directory.

You can configure CDC logging when creating a table or alter an existing table to add the relevant table property.

The `cdc_raw_directory property` in the `cassandra.yaml` file enables you to specify the directory where the database stores the CDC logs. The default locations for this directory are as follows:

Package installations: `/var/lib/Cassandra/cdc_raw`

Tarball installations: `install_location/data/cdc_raw`

Working with Advanced Data Types

You can create several advanced types of data such as the following:

- *Collections*: Enable you to group and store data together in a column.

- *Tuples*: Enable you to store multiple values together in a column.

- *User-Defined Types* (UDTs): Enable you to attach multiple data fields to a column.

Note   An UDT is only for a single keyspace. To use an UDT in a different keyspace, you must create the UDT again.

I explain the three advanced data types in the following sections.

Collections

The collection data type allows you to group data together in a column. For example, if a user has multiple addresses or email addresses, you can store all the addresses together in a collection column.

Remember that a collection is only for data with a limit on the number of values. That is, there's a maximum size of an item in collection. A collection isn't the way to go when storing data that grows without limit, such as events that the database captures and stores every second.

CQL enables you to create three collection types:

- *Set*: A group of elements with unique values. The database doesn't store the values in an ordered fashion, but cqlsh returns the elements in a sorted manner, such as in an alphabetical order for text values.

- *List*: A list groups multiple values, but the values need not be unique. In addition, a list stores elements in a specific order and you can use an index value to insert and query a list's values.

- *Map*: A map establishes a relationship between two items with the help of a key-value pair. For each key, there's a single value, and you can't have duplicates.

Tip   Cassandra reads a whole collection, thus adversely affecting performance. Keep the collections way smaller than the maximum limit for each collection type.

You declare a collection column by specifying the collection type (set, list, or map), followed by a type such as int or text, in angle brackets. Here are a couple of examples:

```
list<text>
list<int>
```

You can create an index on all three types of collection columns.

In the following sections, I show how to create the three collection types.

Creating the Set Type

You use the set type when you have data in a column that has a many-to-one relationship with data in another column. The set type helps in cases like this since you don't join tables in Cassandra.

For example, in the following example, an individual cyclist, represented by the id column, can be a member of multiple teams over time:

```
cqlsh> CREATE TABLE cycling.cyclist_career_teams
      ( id UUID PRIMARY KEY, lastname text, teams set<text);
```

You can query the teams that are part of the set in the following way:

```
cqlsh:cycling> SELECT lastname,teams
          ...               FROM cycling.cyclist_career_teams;
 lastname        | teams
-----------------+---------------------------------------------------------
--------------------------------------------------
    ARMITSTEAD |                         {'AA Drink - Leontien.nl',
              'Boels-Dolmans Cycling Team', 'Team Garmin - Cervelo'}
          VOS | {'Nederland bloeit', 'Rabobank Women Team',
              'Rabobank-Liv Giant', 'Rabobank-Liv Woman Cycling Team'}
        BRAND |   {'AA Drink - Leontien.nl', 'Leontien.nl',
              'Rabobank-Liv Giant', 'Rabobank-Liv Woman Cycling Team'}
```

241

```
VAN DER BREGGEN |                    {'Rabobank-Liv Woman Cycling Team',
                                     'Sengers Ladies Cycling Team', 'Team
                                     Flexpoint'}

(4 rows)
cqlsh:cycling>
```

The maximum size of an item in the set collection type is 65,535 bytes.

Note   You can *expire* each element of a collection by specifying individual TTL properties.

Creating the List Type

Use the list data type when handling columns that have a many-to-many relationship with another column. In the following example, the column events is a list that stores all races during each month:

```
cqlsh> CREATE TABLE cycling.upcoming_calendar ( year int,
        month int, events list<text>,
        PRIMARY KEY ( year, month ));
```

You can query the table for a list of events in a month and year.

```
cqlsh:cycling> SELECT * FROM cycling.upcoming_calendar WHERE year=2015 AND
month=06;

 year | month | events
------+-------+-------------------------------------------------
 2015 |     6 | ['Criterium du Dauphine', 'Tour de Suisse']

(1 rows)
cqlsh:cycling>
```

The maximum size of an item in a list collection type is 2GB.

Creating the Map Type

The following example shows how to create a table with a map type. The column teams has the map type, with each team showing the year and the name of the team a cyclist belonged to during that year.

```
cqlsh> CREATE TABLE cycling.cyclist_teams ( id UUID PRIMARY KEY,
        lastname text,
        firstname text, teams map<int,text> );
```

You can query the table as shown here:

```
cqlsh> SELECT lastname, firstname, teams
        FROM cycling.cyclist_teams;
```

This query shows you all the teams associated with a cyclist for a specific year.

Specifying Frozen Values in Collections

You can denote a type as *frozen* to serialize multiple components into a single value. You can't update individual fields in a frozen type (Cassandra considers the values together as a blob).

Here's an example:

```
cqlsh> CREATE TABLE mykeyspace.users (
        id uuid PRIMARY KEY,
        name frozen <fullname>,
        direct_reports set<frozen <fullname>>,      // a collection set
        addresses map<text, frozen <address>>       // a collection map
        score set<frozen <set<int>>>                // a set with a nested
        frozen set
        );
```

Tuples

You can use the tuple data type to store multiple values together in a column. Here's an example that shows how to create a table with the tuple data type for one of the columns:

```
cqlsh> CREATE TABLE cycling.nation_rank (nation text PRIMARY KEY, info
        tuple<int,text,int>);
```

The following query retrieves data from the `nation_rank` table:

```
cqlsh> SELECT * FROM cycling.nation_rank;
```

User-Defined Types

You can attach multiple named and typed data fields to the same column by defining your own data types, called *user-defined types*. You can specify any valid data type, including collections or even other UDTs, as fields for an UDT.

The following example shows how to create a simple UDT named `cyclist.fullname`:

```
cqlsh> CREATE TYPE cycling.fullname (firstname text, lastname text);
```

You can then create the following table that uses the new type *cycling.fullname*:

```
cqlsh> CREATE TABLE cycling.race_winners
       (race_name text, race_position int, cyclist_name FROZEN<fullname>,
       PRIMARY KEY (race_name, race_position));
```

After inserting data into the table, you can run the following query:

```
cqlsh:cycling> SELECT * FROM cycling.race_winners
               WHERE race_name = 'National Championships South Africa WJ-
               ITT (CN)';
 race_name                                          | race_position
| cyclist_name
----------------------------------------------------+---------------+---------
--------------------------------------------
 National Championships South Africa WJ-ITT (CN) |               1
|     {firstname: 'Frances', lastname: 'DU TOUT'}
 National Championships South Africa WJ-ITT (CN) |               2
|     {firstname: 'Lynette', lastname: 'BENSON'}
...
 (5 rows)
cqlsh:cycling>
```

Notice how the query makes use of the type *full_name* that you defined to show the values of the firstname and the lastname columns for a cyclist.

User-Defined Functions and User-Defined Aggregates

When dealing with large data sets and clusters with hundreds of nodes, there's a dramatic difference between applying aggregations on the client side and pushing huge computations to the server side. By letting the server handle the computations, you'll save network bandwidth, simplify the code on the client side, and reap additional benefits.

In the following sections, I explain the essentials of user-defined functions (UDFs) and user-defined aggregates (UDAs).

User-Defined Functions

The first thing you need to know about an UDF is that its scope isn't database-wide, as is the case with functions and procedures in a database such as Oracle or MySQL. A user-defined function has only a *keyspace-wide* scope.

Things to Remember About UDFs

The following are the key things you need to know about UDFs.

- You can use a language such as Java, JavaScript, Groovy, and Scala to code your UDF.

- You can treat null input in two ways:

  - You can specify the CALLED ON NULL INPUT clause, which means that Cassandra will always call the UDF.

  - If you specify the RETURNS NULL ON NULL INPUT clause, Cassandra skips execution of the UDF and returns a null instead if any of the arguments are null.

- As with the input parameters, you must specify a valid type for the return type, such as a primitive, collection, tuples, or CDT.

Syntax of an UDF

The syntax for creating an UDF is as follows:

```
CREATE [OR REPLACE] FUNCTION [IF NOT EXISTS]
[keyspace.]functionName (param1 type1, param2 type2, ...)
```

245

```
CALLED ON NULL INPUT | RETURNS NULL ON NULL INPUT
RETURN returnType
LANGUAGE language
AS '
        // your source code
';
```

The OR REPLACE clause and the IF NOT EXISTS clause are mutually exclusive.

An Example UDF

The following is an example that shows how to create and use an UDF. The UDF is named maxof (currentvalue int, testvalue int). This function allows you to get the maximum of two integers that you pass to it.

By default, UDFs are disabled. You must therefore first set enable_user_defined_ functions: true to enable in the cassandra.yaml file before trying the following:

```
cqlsh> CREATE FUNCTION maxof(currentvalue int, testvalue int)
       RETURNS NULL ON NULL INPUT
       RETURNS int
       LANGUAGE java
       AS 'return Math.max(currentvalue,testvalue);';
```

You can then create a test table and insert some data into it, following which you can execute a query with your new UDF maxOf to get the maximum value between two values, val1 and val2. Some input values are purposely specified as null. Here is the example:

```
cqlsh> SELECT id,val1,val2,maxOf(val1,val2) FROM test WHERE id IN(1,2,3);
  id | val1 | val2 | udf.maxof(val1, val2)
-----+------+------+-----------------------
   1 |  100 |  200 |                   200
   2 |  100 | null |                  null
   3 | null |  200 |                  null
```

Since the UDF specifies the RETURNS NULL ON NULL INPUT clause, Cassandra returns a null whenever the input argument is null.

User-Defined Aggregate Functions

Cassandra enables you to create user-defined aggregate functions that you can apply to data as part of a query result.

You create the aggregate function first and your query can include only the aggregate function by itself, without any columns.

Built-in Functions and Aggregates

Cassandra offers several handy built-in functions and aggregates you can use. Here's a listing:

- COUNT: The COUNT function gets you a count of the number of rows.

- SUM: The SUM function enables you to add all values for a specific column.

- MIN/MAX: The MIN/MAX functions help you compute the minimum/ maximum values for a column.

- AVG: The AVG function computes the average of all values in a given column.

Summary

Understanding the proper uses of indexing in a Cassandra is tricky since it's so different from the way you use indexing in traditional relational databases.

Cassandra collections such as sets, maps, lists, and UDFs are highly useful and it's good to know how they work.

CHAPTER 7

Cassandra on Docker, Apache Spark, and the Cassandra Cluster Manager

Chapters 2 and 3 showed how to create a generic Cassandra cluster on multiple nodes. Now that you know quite a bit about the architecture of Cassandra, as well as how to configure a cluster, it's time to learn about orchestrating a Cassandra cluster in various environments.

Containers are all the rage now. This chapter first shows you how to create a Cassandra cluster using Docker containers. I quickly explain what Docker containers are and how to work with them. I then show you how to install and configure Cassandra on Docker.

Following the Cassandra on Docker discussion, I show how to create a Cassandra database using Docker-Compose and BDD (behavior-driven development).

The Cassandra Cluster Manager, contrary to what its name might indicate, is not an internal Cassandra tool to manage a cluster. Rather, it's a handy tool for setting up a multi-node Cassandra cluster on a single server. While CCM is for development and testing, it's nice to know how to set up clusters with this tool.

Finally, I wrap up this chapter with a very brief introduction to the emerging technology stack named SMACK (Apache Spark, Apache Mesos, Akka, Cassandra, and Apache Kafka), a set of popular open source tools that is increasingly being used to create distributed enterprise applications.

© Sam R. Alapati 2018
S. R. Alapati, *Expert Apache Cassandra Administration*, https://doi.org/10.1007/978-1-4842-3126-5_7

Cassandra and Docker

In this section, I show how to run a Cassandra cluster on Docker containers. A container is a lightweight, stand-alone, executable package of software that includes everything necessary to run the software, such as the code, runtime, system tools, system libraries, and configurations settings.

Docker: A Quick Introduction

Docker is an application that makes it easy for you run applications within a container. A container is like a virtual machine (VM) but is much more portable. Unlike a VM, which runs in a hypervisor, it runs directly on the host operating system.

Installing Docker

In this section, I show how to install Docker on an Ubuntu 16.04 server.

1. Docker is available as part of the Ubuntu 16.04 installation. However, since this may not be the latest version, download it from the source, shown here:

   ```
   $ curl -fsSL https://download.docker.com/linux/ubuntu/gpg
   | sudo apt-key add -
   OK
   $
   ```

2. Add the Docker repository to the APT source by doing the following:

   ```
   $ sudo add-apt-repository "deb [arch=amd64] https://
   download.docker.com/linux/ubuntu $(lsb_release -cs)
   stable"
   $
   ```

3. Update the package database with the new Docker packages you just added.

   ```
   $ sudo apt-get update
   ```

4. Do the following to make sure that you're installing Docker CE (Community Edition) from the Docker repo, instead of the default Ubuntu Repo:

```
$ apt-cache policy docker-ce
docker-ce:
  Installed: (none)
  Candidate: 17.06.0~ce-0~ubuntu
  Version table:
     17.06.0~ce-0~ubuntu 500
        500 https://download.docker.com/linux/ubuntu xenial/stable
        amd64 Packages
...
$
```

The output verifies that the available Docker install files are for Ubuntu-Xenial (16.04), so you're good to go. Your version number for docker-ce may be different from mine.

5. At this point, you're ready to install Docker.

```
$ sudo apt-get install -y docker-ce
Reading package lists... Done
Building dependency tree
Reading state information... Done
...
Processing triggers for ureadahead (0.100.0-19) ...
$
```

Managing Docker

You can next check that Docker is running via the `systemctl status docker` command.

```
$ sudo systemctl status docker
• docker.service - Docker Application Container Engine
   Loaded: loaded (/lib/systemd/system/docker.service; enabled; vendor
   preset: e
   Active: active (running) since Thu 2017-06-29 10:50:24 EDT; 5s ago
     Docs: https://docs.docker.com
```

```
Main PID: 6598 (dockerd)
   CGroup: /system.slice/docker.service
           ├─6598 /usr/bin/dockerd -H fd://
           └─6603 docker-containerd -l unix:///var/run/docker/
              libcontainerd/dock
...
$
```

The output indicates that the Docker service is running.

Using the Docker Command Line Utility

You manager a Docker instance with the *docker* command line utility. Just type docker at the command line to view all the available options, which are quite a few.

For example, you can view system-wide information about Docker by doing this:

```
$ sudo docker info
Containers: 6
 Running: 0
 Paused: 0
 Stopped: 6
...
Operating System: Ubuntu 16.04.2 LTS
OSType: linux
Architecture: x86_64
CPUs: 1
Total Memory: 7.796GiB
Name: ubuntu1
...
$
```

Understanding Docker Images

All Dockers containers run from a docker image. By default, Docker pulls its images from Docker Hub, a public Docker Registry maintained by the Docker company. Others, including Linux distributions, also host their images on the Docker Hub.

You can check if your Docker installation can access and download images from Docker Hub by running the following command:

```
$ sudo docker run hello-world
Hello from Docker!
This message shows that your installation appears to be working correctly.
$
```

You can search for images on the Docker Hub with the `docker search` command. If you want to look for all available Ubuntu images, type this:

```
$ sudo docker search ubuntu
```

You can then download the image you want by specifying the name of the image with the `docker pull` command.

```
$ sudo docker pull ubuntu
```

You can view all the images you've downloaded to your server by doing this:

```
$ sudo docker images
REPOSITORY        TAG           IMAGE ID          CREATED         SIZE
hello-world       latest        1815c82652c0      2 weeks ago     1.84kB
ubuntu            latest        d355ed3537e9      12 days ago     119MB
...
$
```

And finally, you can run a container using the images you've downloaded with the `docker run` command.

```
$ sudo docker run -it ubuntu
samalapati@bd78ccf18941:/#
```

This will start a container running the Ubuntu OS. The i option stands for interactive and the t option offers shell access into the container. Note that the command prompt changes since you're now inside the new Ubuntu container. The container id is bd78ccf18941.

Now that you have the basics of Docker under your belt, it's time to learn about how you can run Cassandra containers on Docker.

Running a Cassandra Cluster on Docker

Let's create a three-node Cassandra cluster on Docker. Start the first Cassandra instance by having Docker pull the Cassandra 3.04 image from Docker Hub.

```
$ sudo docker run --name cassandra1 -m 4g -d cassandra:3.0.4
34842881b8d4b42a9cc6627405f6dd63a21cace23276091e869e8b708fb6ce4b
$
```

In this command,

- The name option enables you to specify the name for this instance.

- The -m option lets you specify the memory for this instance.

- The cassandra:3.0.4 clause tells Docker the version it should download for the Cassandra image.

You next run the docker ps command to check if the Cassandra container instance *cassandra1* is running.

```
$ sudo docker ps
CONTAINER ID        IMAGE           COMMAND                 CREATED
STATUS              PORTS                                   NAMES
34842881b8d4        cassandra:3.0.4     "/docker-entrypoin..."   8 seconds
ago      Up 7 seconds            7000-7001/tcp, 7199/tcp, 9042/tcp, 9160/
tcp    cassandra1
#
```

You run the familiar nodetool status command to check the status of the Cassandra node *cassandra1*.

```
$ nodetool status
nodetool: command not found
samalapati@ubuntu1:/tmp# sudo docker exec -i -t cassandra1 sh -c 'nodetool
status'
Datacenter: datacenter1
=======================
Status=Up/Down
```

```
|/ State=Normal/Leaving/Joining/Moving
-- Address     Load      Tokens      Owns (effective)  Host ID      Rack
UN  172.17.0.2  102.52 KB  256        100.0%      f217c613-3eb9-445a-884a-
183e8d177927   rack1
root@ubuntu1:/tmp#
```

Now that the first Cassandra node is running, it's time to create a cluster by creating a second node, using the first node as the seed node. The second node and any other nodes you create will gossip to this node and register themselves. To do this, you need the IP of your first Cassandra node, cassandra1. You can get it by doing this:

```
$ sudo docker inspect -f{{.NetworkSettings.IPAddress}}' cassandra1
172.17.0.3
$
```

The command reveals that the IP address of the first node is 172.17.0.3. Now you have everything you need to create the second Casandra node, which you name cassandra2.

```
# docker run --name cassandra2 -m 2g -d -e CASSANDRA_SEEDS="172.17.0.3"
cassandra:3.0.4
30ed76a87b3c0f2f63075559817a2f125fc250731c75416639e446aa9f0b7774
#
```

You finally add the third node to the Cassandra cluster, named cassandra3.

```
# docker run --name cassandra3 -m 2g -d -e CASSANDRA_SEEDS="172.17.0.3"
cassandra:3.0.4
7528dc5679b02a59bb13f30ce0dc8c7617f0c268fd52256d8e2d9ea5faf0586a
#
```

The CASSANDRA_SEEDS environment variable is a comma-separated list of IP addresses that the gossip protocol uses to bootstrap the new Cassandra nodes you create when they join the cluster.

Check the Docker processes to ensure that all three Cassandra containers are running.

```
# sudo docker ps
CONTAINER ID  IMAGE         COMMAND        CREATED       STATUS    PORTS          NAMES
7528dc5679b0        cassandra:3.0.4     "/docker-entrypoin..."   5 seconds
ago   Up 4 seconds      7000-7001/tcp, 7199/tcp, 9042/tcp, 9160/tcp    cass3
30ed76a87b3c        cassandra:3.0.4     "/docker-entrypoin..."   5 minutes ago
Up 5 minutes          7000-7001/tcp, 7199/tcp, 9042/tcp, 9160/tcp         cass2
34842881b8d4        cassandra:3.0.4     "/docker-entrypoin..."   8 minutes ago
Up 8 minutes          7000-7001/tcp, 7199/tcp, 9042/tcp, 9160/tcp    cassandra1
#
```

A final nodetool status command shows that all three nodes of the Cassandra cluster are running.

```
# sudo docker exec -i -t cassandra1 sh -c 'nodetool status'
Datacenter: datacenter1
========================
Status=Up/Down
|/ State=Normal/Leaving/Joining/Moving
--  Address      Load       Tokens      Owns (effective)   Host ID       Rack
UN  172.17.0.3  101.89 KB   256         67.3%              968690b8-8dc4-
4242-943f-b62d37e44084   rack1
UN  172.17.0.2  107.45 KB   256         68.7%              f217c613-3eb9-
445a-884a-183e8d177927   rack1
UN  172.17.0.4  15.42 KB    256         64.0%              820f24f7-d931-
4a9f-87e3-8abd72d8a9bb   rack1
#
```

Note that you can't run the Cassandra utilities nodetool, cqlsh, or any other such commands in the main Linux server. Since Cassandra is running inside the Docker container, you can run these commands only in Docker. You must use the docker exec command to run commands inside a Docker container.

Running cqlsh in the Docker-Based Cluster

To run cqlsh from one of the three nodes in the Docker-based cluster, you need to "bash" into one of the nodes, say the node cassandra1, as shown here:

```
$ sudo docker exec -it cassandra1 /bin/bash
samalapati@34842881b8d4:/# cqlsh
Connected to Test Cluster at 127.0.0.1:9042.
[cqlsh 5.0.1 | Cassandra 3.0.4 | CQL spec 3.4.0 | Native protocol v4]
Use HELP for help.
cqlsh>
```

The docker exec command enables you to run commands inside a Docker container. The previous command provides you a bash shell within the container cassandra1.

Once you log into the node through the bash shell, run the cqlsh command as usual to get to the cqlsh prompt.

Alternatively, you can specify the Docker link option to connect to a Docker instance and log into cqlsh with the help of the exec cqlsh <IPAddress> option.

```
# docker run -it --link cassandra1 --rm cassandra:3.0.4 \
> sh -c 'exec cqlsh 172.17.0.2'
Connected to Test Cluster at 172.17.0.2:9042.
[cqlsh 5.0.1 | Cassandra 3.0.4 | CQL spec 3.4.0 | Native protocol v4]
Use HELP for help.
cqlsh>
```

Setting Cassandra Environment Variables for Docker

You can pass one of several environment variables that are available to you when you start a Cassandra image on Docker. More specifically, you pass these environment variables on the command line when you issue the docker run command. Each of these environment variables that you pass at the command line sets the value of a related variable in the cassandra.yaml file of the node.

You learned how to specify the CASSANDRA_SEEDS environment variable in the previous section. In addition to that variable, there are several others. Here are the most useful:

- CASSANDRA_LISTEN_ADDRESS: Determines the IP address to listen for from incoming connections. If you choose the default value of auto, it sets the listen_address option in the cassandra.yaml file to the IP address of the container.

- CASSANDRA_BROADCAST_ADDRESS: This variable specifies the IP address the node advertises to the rest of the nodes in the cluster. This will set the value of the broadcast_address and the broadcast_rpc_address properties in the cassandra.yaml file.

- CASSANDRA_ENDPOINT_SNITCH: This property enables you to set the snitch implementation via setting the endpoint_snitch property in the cassandra.yaml file.

- CASSANDRA_NUM_TOKENS: This variable sets the number of tokens for this node by setting the num_tokens property.

- CASSANDRA_DC: Sets the name of the datacenter for this node by setting the value of the dc property. If you omit this variable, the name of the data center defaults to *datacenter1*.

- CASSANDRA_RACK: This variable sets the name of the rack by setting the value of the rack property. If you omit it, as I did here, the variable defaults to *rack1*.

Storing Cassandra Data on Docker

There's more than one way to store data used by Cassandra that runs in a Docker container. Two of the options are the following:

- *Using Docker volumes*: By default, Docker manages the data storage by writing the Cassandra data files to the host's file system using internal volume management. While the containers can easily view and access this data, the same isn't true for tools and applications that run on Docker, but directly on the host server. They may not be able to easily locate Cassandra's data files. Also, when you shut down the container, or it dies, or if the Docker host dies, the data is lost.

- *Creating a data directory on the host server*: You can also mount a host directory as a data volume. You mount the data directory on the host to a directory visible from inside a container. This way, tools and applications can easily access the files in the host system. You just need to make sure to correctly set up the directory permissions and any ACLs (access control lists, which control access to directories), etc. on the host system so users can access those directories.

Here's an example that shows how to mount a volume with the Docker host's file system:

```
$ docker run –name my_container -v /host/dir://container/dir cassandra
```

When you do this, everything that the container process writes to the `/container/dir` directory is written directly to the `/host/dir` directory on the host file system.

Note   CassandraCloud is a tool from Cloudurable that simplifies the configuration of Cassandra for various environments, including, EC2, Docker, and VirtualBox. For details, please go to `https://cloudurable.github.io/cassandra-cloud/`.

Creating a Cassandra Cluster Using Docker-Compose and Behavior-Driven Development

You can quickly set development and test environments for Cassandra by using behavior-driven development. BDD is in many ways similar to test-driven development (TDD), but whereas the latter focuses on unit testing and integration testing, TDD focuses on the interaction between a system and its users.

When you use BDD, you describe what should happen in various scenarios. For example, for a web site, you describe the workflow in the form of user clicks and user entry of data into forms. For a system administrator, the workflow is comprised of the commands entered by the system administrator when he or she interacts with systems and programs.

In this section, I show how to use BDD and a tool called Cucumber along with Docker Compose to develop new Docker containers running Cassandra. Here's how BDD and Docker Compose help you:

- BDD helps develop the new Docker containers in a test-driven fashion.

- Docker Compose helps you orchestrate Docker containers.

Together, BDD (plus Cucumber) and Docker-Compose make for an exciting way to spin up new clusters. Although you can do similar things with Vagrant, a tool that makes it easy to spin up virtual machines for development, the method I describe here is far easier and more fun.

Getting the Prerequisites Out of the Way

The three things you need to have to set up a Cassandra cluster using BDD are the following:

- *Docker*: As explained, Docker is a container platform that helps you create and manage containers.

- *Docker Compose*: This is a tool that helps you define and run multi-container Docker applications. You use a Compose file to configure the services your application requires. Following this you can create and start all the services with a single command, as shown here:

  ```
  $ docker-compose up -d
  ```

- *Cucumber*: Cucumber is a testing tool that helps test other software by running automated acceptance tests written in a BDD style.

Installing Docker

You need Docker, and I assume you already have it, if you've followed my earlier discussion on running Docker-based Cassandra clusters. If not, you can download and install Docker.

Installing Docker-Compose

You can install Docker-Compose by doing the following (on an Ubuntu system):

```
# curl -L https://github.com/docker/compose/releases/download/1.14.0/
docker-compose-'uname -s'-'uname -m' > /usr/local/bin/docker-compose
# chmod +x /usr/local/bin/docker-compose
```

Test the installation by invoking docker-compose.

```
# docker-compose
Define and run multi-container applications with Docker.
Usage:
  docker-compose [-f <arg>...] [options] [COMMAND] [ARGS...]
  docker-compose -h|--help
#
```

Installing Cucumber

I'm using an Ubuntu system, so I install Cucumber by doing the following:

```
$ gem install cucumber
Fetching: gherkin-4.1.3.gem (100%)
...
Done installing documentation for gherkin, cucumber-core, cucumber-wire,
cucumber after 4 seconds
4 gems installed
$
```

It's a good idea to install an additional tool called *bundler*. Bundler offers a consistent environment for Ruby-based projects (such as Cucumber, which we're going to use). It does this by tracing and installing all the gems that you need with the correct versions. In other words, bundler offers a way out of dependency hell.

```
$ gem install bundler
Successfully installed bundler-1.15.1
1 gem installed
#
```

Now that you have Docker, Docker-Compose, and Cucumber installed, you're ready to spin up a Cassandra cluster using the BDD approach. But first, a bit about user stories and BDD.

User Stories

BDD uses stories to express what you want to accomplish. It uses plain language-based Gherkin syntax, which is quite similar to the way we naturally speak to describe what you want to do.

You describe your main goals as *features*, as shown here, where you want to deploy a new Cassandra cluster:

> *Feature: Cassandra Cluster*
>
> *As a Big Data Administrator*
>
> *I want to deploy a Cassandra cluster*
>
> *So I can store and process massive quantities of data*

Once you describe your high-level goals, you describe specific scenarios that make the goal happen. In this case, you want to spin up a Cassandra database and test its version.

> *Scenario: Launch a CQL Shell*
>
> *Given the services are running*
>
> *And I run "cqlsh -e 'show version'" on "cassandra"*
>
> *Then I should see "CQL spec 3.4.2"*
>
> *And I should see "Cassandra 3.7"*

Getting Ready to Run the BDD Tests

At this point, instead of doing everything from scratch, you can download the tests and the steps for your work from GitHub. On GitHub, coshxlabs (`www.coshx.com`) has a repository (`https://github.com/coshx/docker-bdd`) that demonstrates how to use BDD and Cucumber to orchestrate Docker containers to help you quickly spin up development environments.

1. You can check out Coshxlabs's code from GitHub.

    ```
    $ git clone https://github.com/coshx/docker-bdd
    $ git reset --hard e9f807c
    ```

2. Move to the directory you cloned from GitHub, which is named
    docker-bdd. This directory contains all the artifacts that you need
    to generate the Docker containers that'll run Cassandra.

    ```
    # cd docker-bdd
    # ls
    circle.yml  docker-compose.yml  example    Gemfile
    README.md
    deploy.sh   dockerfiles              features  Gemfile.lock
    #
    ```

3. Go to the features directory.

    ```
    # cd features
    # ls
    android_app.feature  app.feature  db.feature  step_
    definitions     support  web.feature
    ```

4. Create a file named cassandra.feature and add the Cassandra-
    related feature and scenario described in the previous section.
    Save the file. Here's what the contents of cassandra.feature file
    should look like:

    ```
    # cat cassandnra.feature
    Feature: Cassandra Cluster
      As a Big Data Administrator
      I want to deploy a cassandra cluster
      So I can store and query lots of data

    Scenario: Launch a CQL Shell
      Given the services are running
        And I run "cqlsh cassandra -e 'show version'" on
        "cassandra-dev"
       Then I should see "CQL spec"
        And I should see "Cassandra 3.7"
    ```

5. Change the working directory to the docker-bdd directory, and
    edit the docker-compose.yml file so it contains the following
    lines (the docker-compose.yml file contains the names of the
    images that Docker should download and install).

```
cassandra:
   image: cassandra
```

```
cassandra-dev:
   image: cassandra
   links:
     - cassandra
```

This snippet of code creates two separate Docker containers,
both using the same Docker image ("cassandra"). You need
two containers: one for running the Cassandra instance in the
background and the other for running cqlsh. Remember that your
scenario includes the following:

```
"cqlsh -e 'show version'" on "cassandra"
```

Running the BDD Tests

You're all set now to run the BDD tests with Cucumber. Here's how you do this:

```
# cucumber features/cassandra.feature
Feature: Cassandra Cluster
  As a Big Data Administrator
  I want to deploy a cassandra cluster
  So I can store and query lots of data

cassandra uses an image, skipping
cassandra-dev uses an image, skipping
Pulling cassandra (cassandra:latest)...
latest: Pulling from library/cassandra
ockerbdd_cassandra_1 is up-to-date
dockerbdd_cassandra-dev_1 is up-to-date
  Scenario: Launch a CQL Shell
  # features/cassandra.feature:6
```

```
   Given the services are running
   # features/step_definitions/docker_compose_steps.rb:3
     Running: docker-compose build && (docker-compose up -d || true)
     Process exited successfully
[cqlsh 5.0.1 | Cassandra 3.11.0 | CQL spec 3.4.4 | Native protocol v4]
   And I run "cqlsh cassandra -e 'show version'" on "cassandra-dev"
   # features/step_definitions/docker_compose_steps.rb:14
     Running: docker-compose run cassandra-dev bash -i -c "sleep 1;
     cqlsh cassandra -e 'show version'"
     Process exited successfully
   Then I should see "CQL spec"
   # features/cassandra.feature:9
   And I should see "Cassandra 3.7"
   # features/cassandra.feature:10

1 scenario (1 undefined)
4 steps (2 undefined, 2 passed)
0m8.169s
#
```

Docker-Compose creates the single-node Cassandra cluster for you and BDD runs the tests you specified on this cluster.

The first time you run the Cucumber command shown here, based on the scenario that you specified ("Given the services are running") Docker-Compose builds all the Docker images that you've specified in the docker-compose.yml file. On subsequent runs, docker-compose is aware that both the *cassandra* and the *cassandra-dev* instances use an available image, so it doesn't build the images. Instead, it just starts up the two Cassandra instances.

You can verify that the two Docker containers are running with the docker ps command.

```
$ sudo docker ps
CONTAINER ID    IMAGE      COMMAND       CREATED   STATUS      PORTS      NAMES
b005b61706ca    cassandra   "/docker-entrypoint.s"   44 minutes ago    Up 40
minutes   7000-7001/tcp, 7199/tcp, 9042/tcp, 9160/tcpdockerbdd_cassandra-dev_1
ed526bd724c7        cassandra              "/docker-entrypoint.s"   44 minutes
ago     Up 40 minutes        7000-7001/tcp, 7199/tcp, 9042/tcp, 9160/tcp
dockerbdd_cassandra_1
$
```

And you can bash into a container (dockerbdd_cassandra_dev1) as shown here:

```
$ sudo docker exec -it b005b61706ca bash
/# hostname
b005b61706ca
/#
```

And finally, you can use the CQL shell in the new Docker container as shown here:

```
$ cqlsh
Connected to Test Cluster at 127.0.0.1:9042.
[cqlsh 5.0.1 | Cassandra 3.11.0 | CQL spec 3.4.4 | Native protocol v4]
Use HELP for help.
cqlsh>
```

You can get more details about BDD-driven Cassandra clusters, including details about creating a multi-node Cassandra cluster, from an article written by Coshxlab's Ben Taitelbaum, available at the following location:

www.coshx.com/blog/2016/08/01/cassandra-cluster-in-docker-using-bdd/

Using the Cassandra Cluster Manager to Spin Up Clusters

The Cassandra Cluster Manager (CCM) is a script library that helps you create and run a Cassandra cluster on a local server. CCM isn't for production usage or for stress testing a cluster; it's a way to test and learn how to run a Cassandra cluster. CCM is an alternative to both Docker and Vagrant as a way to quickly spin up a Cassandra cluster.

You use CCM in the following situations:

- When you want to run Cassandra without bothering to install it

- To test failure scenarios

- When developing with multiple versions without worrying about the hardware, since you can run everything on the local node

- For integration testing

- For testing upgrades

What's amazing about CCM is how quickly it enables you to get a new cluster up and running. You can spin up a multi-node (local) cluster in under a minute.

Installing CCM

Installing CCM is easy: just make sure you have Python (at least the 2.7 version) and *ant* installed before you commence the installation of CCM.

1. To install CCM, you need to get the goodies from GitHub by running the following command:

```
root@ubuntu2:/home/samalapati# git clone https://github.
com/pcmanus/ccm.git
Cloning into 'ccm'...
remote: Counting objects: 4372, done.
remote: Compressing objects: 100% (15/15), done.
remote: Total 4372 (delta 8), reused 14 (delta 6),
pack-reused 4351
Receiving objects: 100% (4372/4372), 1.80 MiB | 0 bytes/s,
done.
Resolving deltas: 100% (3040/3040), done.
Checking connectivity... done.
root@ubuntu2:/home/samalapati# ls
```

2. Once you download the scripts and other files from GitHub, run the CCM installation script.

```
$ sudo ./setup.py install
```

The installation files are under the ccm directory.

```
$ ls
ccm        Documents  examples.desktop  Pictures  Templates
Desktop    Downloads  Music             Public    Videos
$ cc ccm
$ ls
ccm  ccmlib  license.txt  MANIFEST.in  misc  README.md
setup.py  ssl  tests
```

```
$ sudo ./setup.py install
running install
running build
running build_py
creating build
...
creating build/scripts-2.7
...
running install_egg_info
$
```

You can check out the various things you can do with CCM by typing ccm, ccm -help, or ccm <command> -h. Of course, the first and most important thing you can do is to create a cluster, so I turn to that task next.

Creating a Cassandra Cluster with CCM

After you install CCM as shown in the previous section, it's time to create a Cassandra cluster. Let's create a three-node cluster with the Cassandra 3.1 release. By default, CCM creates single-token nodes, so tell it to use vnodes.

CCM downloads all the source files it needs and compiles them for you.

```
$ sudo ccm create -n 3 -v 3.1 testcluster --vnodes
18:55:53,875 ccm INFO Downloading http://archive.apache.org/dist/
cassandra/3.1/apache-cassandra-3.1-bin.tar.gz to /tmp/ccm-tlSojg.tar.gz
(29.376MB)
  306318:55:59,731 ccm INFO Extracting /tmp/ccm-tlSojg.tar.gz as
  version 3.1 ...
  30802666  [100.00%]Current cluster is now: testclusster
$
```

The command ccm list shows all the Cassandra clusters on this node that are managed by CCM.

```
$ sudo ccm list
 *testcluster
$
```

Checking the Status of the Cluster

You can check the status of the new three-node cluster thus (note that you can't use the nodetool commands in a CCM environment!):

```
$ sudo ccm status
Cluster: 'testcluster'
----------------------
node1: DOWN (Not initialized)
node3: DOWN (Not initialized)
node2: DOWN (Not initialized)
$
```

All three nodes in the new cluster are DOWN since you created the cluster but haven't started it up.

Starting the Cluster

You execute the ccm start command to start a CCM cluster. In this example, I start the cluster up and check its status with the ccm status command.

```
$ sudo ccm start
$ sudo ccm status
Cluster: 'testcluster'
----------------------
node1: UP
node3: UP
node2: UP
$
```

The command ccm node1 status provides even more details about the status of the cluster.

```
$ sudo node1 status

Datacenter: datacenter1
=======================
Status=Up/Down
```

```
|/ State=Normal/Leaving/Joining/Moving
--  Address     Load      Tokens      Owns      Host ID                Rack
UN  127.0.0.1   86.51 KB  256         ?         659b06e0-a4d9-49f5-b923-
0bab9833daa3  rack1
UN  127.0.0.2   86.47 KB  256         ?         4035e764-a152-40e6-ac9e-
f9ddf1e603b4  rack1
UN  127.0.0.3   92.34 KB  256         ?         40182f69-a81f-4a23-90ae-
e9e1f0929c5e  rack1
Note: Non-system keyspaces don't have the same replication settings,
effective ownership information is meaningless
$
```

You can stop the cluster nodes by running the command ccm stop.

Working with CCM

In this section, I offer a glimpse into CCM by showing how to run a few important commands, the first of which allows you access to the CQL shell.

Using cqlsh with CCM

To work with CQL, run the ccm command with the cqlsh option, making sure you specify a node.

```
$ sudo  ccm node1 cqlsh
Connected to testcluster2 at 127.0.0.1:9042.
[cqlsh 5.0.1 | Cassandra 3.11.0 | CQL spec 3.4.4 | Native protocol v4]
Use HELP for help.
cqlsh> cqlsh
```

Get SSTable Information

The ccm node1 getsstables command shows all SSTables on this node.

```
$ sudo ccm node1 getsstables
/samalapati/.ccm/testcluster/node1/data0/system_schema/triggers-4df70b666b0
5325195a132b54005fd48/ma-1-big-Data.db
```

```
/samalapati/.ccm/testcluster/node1/data0/system_schema/types-5a8b1ca866023f
77a0459273d308917a/ma-1-big-Data.db
...
$
```

Running Apache Spark with Cassandra

Apache Spark is an immensely popular, fast general engine for large-scale data processing. Spark runs on Apache Hadoop, Apache Mesos, in the cloud, or standalone. Spark can access data sources such as Hadoop's HDFS, Hbase, AWS's S3, and of course, Apache Cassandra.

In this section, I show how to install Apache Spark and work with a Cassandra database from a Spark cluster through DataStax's Spark-Casandra connector.

Installing the Prerequisites

To install Apache Spark, there are a couple of prerequisites, which I review in this section. Make sure that you either already have the following on the server where you're going to run the Spark cluster or install them if they're missing.

1. Make sure you have at least Java 1.8.x installed.

    ```
    # java -version
    java version "1.8.0_131"
    Java(TM) SE Runtime Environment (build 1.8.0_131-b11)
    Java HotSpot(TM) 64-Bit Server VM (build 25.131-b11,
    mixed mode)
    #
    ```

2. You'll also need Scala.

    ```
    # scala -version
    Scala code runner version 2.11.7 -- Copyright 2002-2013,
    LAMP/EPFL
    #
    ```

    You can install Scala by doing the following:

    ```
    # wget www.scala-lang.org/files/archive/scala-2.11.8.deb
    # sudo dpkg -i scala-2.11.8.deb
    ```

3. Finally, you need SBT (Simple Build Tool) to compile the Scala programs.

    ```
    # echo "deb https://dl.bintray.com/sbt/debian /" | sudo
    tee -a /etc/apt/sources.list.d/sbt.list
    # sudo apt-key adv --keyserver hkp://keyserver.ubuntu.
    com:80 --recv 2EE0EA64E40A89B84B2DF73499E82A75642AC823
    # sudo apt-get update
    # sudo apt-get install sbt
    ```

Check the version of SBT:

```
# sbt sbt-version
...
 [info] 0.13.15
#
```

It's very important that you have the correct SBT version; incompatible versions are the source of a lot of headaches when working with Spark and Cassandra.

Now you're ready to install Apache Spark.

Installing Apache Spark

Installing and getting going with Apache Spark for your purpose here, which is to learn how to connect to a Cassandra cluster from a Spark cluster, is very straightforward. Follow these steps to install Spark.

1. `$ wget http://d3kbcqa49mib13.cloudfront.net/spark-2.0.2-bin-hadoop2.7.tgz`

2. `$ tar zxf spark-2.0.2-bin-hadoop2.7.tgz`

3. `$ sudo mv spark-2.0.2-bin-hadoop2.7/usr/local/spark/`

4. Once you've installed the Spark binaries, set a home directory for Spark.

    `export $SPARK_HOME= /usr/local/spark`

5. Export the PATH variable.

```
export PATH="/usr/local/sbin:/usr/local/bin:/usr/sbin:/usr/
bin:/sbin:/bin:/usr/games:/usr/local/games:$SPARK_HOME/bin"
```

Configuring the Spark Cluster

You can run a Spark cluster by managing it with the YARN cluster manager (Hadoop), Apache Mesos, or a Spark standalone cluster. Let's try the last alternative so you can run Spark alongside the Cassandra cluster.

Spark consists of a single master node and multiple worker nodes. For simplicity, let's run both the master and worker services on the same node, but it's easy to configure the workers to run on different nodes.

The main configuration file is the spark-env file, located in the /etc/spark/conf directory. In this file, you specify the options for the daemons used in the standalone deployment mode for Spark. That is, you specify the options for the master and the worker nodes.

You can leave all the configuration attributes such as SPARK_WORKER_CORES and SPARK_WORKER_MEMORY at their default values. Just make sure that you specify the details for the node where the Spark master instance will run.

```
export SPARK_MASTER_HOST=192.168.159.129
```

This way, the master will run on the server with the IP 192.168.159.129. You must add this property to the /etc/spark/conf/spark-env file on each of the nodes in the cluster. Since you're running both the master and the worker instances on the same node, you're good.

In a production cluster, you can configure Spark to run as a service, but here I'm concerned only with demonstrating how to connect to Cassandra from Spark, so we won't bother with creating a Spark service. Instead, let's manually start and stop the Spark cluster.

Starting Up the Spark Cluster

Since the Spark cluster has just one worker instance running along with the master instance on the same node, here's what you need to do to start up the Spark cluster.

1. Move to the sbin directory of Spark, which has the start/stop scripts for the cluster.

```
# cd /usr/local/spark/spark-2.0.2-bin-hadoop2.7/sbin
```

2. Start the Spark master instance.

```
# ./start-master.sh
starting org.apache.spark.deploy.master.Master, logging
to /usr/local/spark/spark-2.0.2-bin-hadoop2.7/logs/spark-
root-org.apache.spark.deploy.master.Master-1-ubuntu.out
#
```

3. Start the single worker instance.

```
# ./start-slave.sh spark://ubuntu:7077
starting org.apache.spark.deploy.worker.Worker, logging to /usr/
local/spark/spark-2.0.2-bin-hadoop2.7/logs/spark-root-org.apache.
spark.deploy.worker.Worker-1-ubuntu.out
#
```

4. Make sure that the master and worker instances are up and
    running.

```
# ps -ef|grep spark
root      14941    2008  0 10:37 pts/1     00:00:04 /usr/
lib/jvm/java-8-oracle/jre/bin/java -cp /usr/local/spark/
spark-2.0.2-bin-hadoop2.7/conf/:/usr/local/spark/spark-
2.0.2-bin-hadoop2.7/jars/* -Xmx1g org.apache.spark.
deploy.master.Master --host ubuntu --port 7077 --webui-
port 8080
root      15073    2008  1 10:39 pts/1     00:00:04 /
usr/lib/jvm/java-8-oracle/jre/bin/java -cp /usr/local/
spark/spark-2.0.2-bin-hadoop2.7/conf/:/usr/local/spark/
spark-2.0.2-bin-hadoop2.7/jars/* -Xmx1g org.apache.
spark.deploy.worker.Worker --webui-port 8081 spark://
ubuntu:7077
#
```

Everything looks right thus far, and the Spark cluster is up and running. In the next
section, I show how to connect to a Cassandra database from the Spark cluster.

Connecting to Cassandra from a Spark Cluster

One way to access your Cassandra tables in Spark is to transform them into Spark RDDs (Resilient Distributed Dataset). You can view a Cassandra's table data as a Spark RDD, and you can write a Spark RDD to a Cassandra table.

You can access Cassandra from Spark in several ways. The following projects provide the libraries to enable you to read and write to Cassandra tables from a Spark program:

- https://github.com/datastax/spark-cassandra-connector

- http://tuplejump.github.io/calliope/pyspark.html

- https://github.com/TargetHolding/pyspark-cassandra

Let's use the first of these projects, which is the *spark-cassandra-connector*.

The *spark-cassandra-connector (*DataStax Cassandra connector*)*, a Scala library offered by DataStax, makes it possible to work with Spark and Cassandra together. You can get the latest version of this connector from https://github.com/datastax/spark-cassandra-connector.

When you write code to access Cassandra from Spark, your code, along with the *spark-Cassandra-connector* and all the necessary dependencies, are packaged and sent to the Spark cluster's nodes.

You submit jobs to a Spark cluster through the spark-submit interface. When you do this, you can include a fat jar at the command line to include the *spark-cassandra connector*.

The example here will show how to perform ad hoc work from a Spark cluster. So use spark-shell instead of spark-submit to interact with Spark (and Cassandra).

In the following example, you start spark-shell and invoke the *spark-cassandra-connector* so you can communicate with Cassandra from the Spark cluster. The –packages option enables you to specify the spark-cassandra-connector. This option downloads the connector and its dependencies from the Spark packages.

Note that you need to run the spark-shell command from the $SPARK_HOME/bin directory and not from the $SPARK_HOME/sbin directory, from where you start the Spark instances.

```
bin# ./spark-shell --conf spark.cassandra.connection.host=192.168.159.129
--packages datastax:spark-cassandra-connector:2.0.0-M2-s_2.11
...
```

```
org.scala-lang#scala-reflect;2.11.8 from central in [default]
---------------------------------------------------------------------
|                 |           modules        ||   artifacts    |
|      conf       | number| search|dwnlded|evicted|| number|dwnlded|
---------------------------------------------------------------------
|     default     |   7   |   1   |   1   |   0   ||   7   |   0   |
---------------------------------------------------------------------
```

Using Spark's default log4j profile: org/apache/spark/log4j-defaults.properties
Setting default log level to "WARN".Spark context Web UI available at
http://192.168.159.129:4040
Spark context available as 'sc' (master = local[*], app id =
local-1499634822206).
Spark session available as 'spark'.
Welcome to

```
      ____              __
     / __/__  ___ _____/ /__
    _\ \/ _ \/ _ `/ __/  '_/
   /___/ .__/\_,_/_/ /_/\_\   version 2.0.2
      /_/
```

Using Scala version 2.11.8 (Java HotSpot(TM) 64-Bit Server VM, Java
1.8.0_131)
Type in expressions to have them evaluated.
Type :help for more information.
scala>

You can use either Scala or Python (or Java) to write Spark code. In this case, you chose Scala by executing the spark-shell command. To code in Python, run the pyspark command instead, as shown here:

```
root@ubuntu:/usr/local/spark/spark-2.0.2-bin-hadoop2.7/bin# ./pyspark \
> --master spark://ubuntu:7077 \
> --packages datastax:spark-cassandra-connector:2.0.0-M2-s_2.11 \
> --conf spark.cassandra.connection.host=ubuntu
Python 2.7.12 (default, Nov 19 2016, 06:48:10)
...
```

```
Welcome to
   / __/__  ___ _____/ /__
  _\ \/ _ \/ _ `/ __/  '_/
 /__ / .__/\_,_/_/ /_/\_\   version 2.0.2
    /_/
Using Python version 2.7.12 (default, Nov 19 2016 06:48:10)
SparkSession available as 'spark'.
>>> posts = spark.read.format("org.apache.spark.sql.cassandra").
options(table="posts", keyspace="posts_db").load()
```

Working with Cassandra from Spark

Before you can run commands to access a Cassandra database, you must first import the Spark connector namespace, as shown here:

```
scala> import com.datastax.spark.connector._
import com.datastax.spark.connector._
scala>
```

Let's work with the cyclist_name table from the *cyclist* keyspace. The first thing to do is to create a Spark RDD for this table's data.

```
scala> val rdd1 = sc.cassandraTable("cycling", "cyclist_name")
rdd: com.datastax.spark.connector.rdd.CassandraTableScanRDD[com.datastax.
spark.connector.CassandraRow] = CassandraTableScanRDD[2] at RDD at
CassandraRDD.scala:18
scala>
```

In the command, sc stands for Spark Context. Spark automatically creates a Spark Context named sc when you invoke spark-shell. Spark Context is the main entry point for all of Spark's functionality, and you must use it in the way I showed you here.

Spark Context enables you to connect to a spark cluster, and you can use it create Spark RDDs and dataframes, datasets, etc. in a Spark cluster. In a way, Spark Context is the Spark cluster's coordinator (the driver program). The SparkContext object coordinates the independent sets of processes in a cluster that constitute a Spark application. Spark Context connects to the cluster manager, which could be Mesos, YARN, or a standalone cluster manager (as mentioned earlier, I'm using the Spark standalone cluster).

Next, let's test the Spark-cluster-to-Cassandra-database connection by running the following code, which uses the function `first` to get the first row in the table CYCLIST_NAME:

```scala
scala> println(rdd.first)
CassandraRow{id: e7ae5cf3-d358-4d99-b900-85902fda9bb0, firstname: Alex,
lastname: FRAME}
scala>
```

The following command uses the `foreach` construct to print all rows in the table CYCLIST_NAME:

```scala
scala> rdd.foreach(println)
CassandraRow{id: e7ae5cf3-d358-4d99-b900-85902fda9bb0, firstname: Alex,
lastname: FRAME}
CassandraRow{id: 5b6962dd-3f90-4c93-8f61-eabfa4a803e2, firstname: Marianne,
lastname: VOS}
CassandraRow{id: e7cd5752-bc0d-4157-a80f-7523add8dbcd, firstname: Anna,
lastname: VAN DER BREGGEN}
scala>
```

Now that you've successfully tested that you can retrieve data from a Cassandra table, let's insert some data into a Cassandra table from Spark. You first create an RDD named *collection*.

```scala
scala> val collection = sc.parallelize(Seq(("cat", 30), ("fox", 40)))
collection: org.apache.spark.rdd.RDD[(String, Int)] =
ParallelCollectionRDD[4] at parallelize at <console>:27
scala>
```

Following this, you execute Spark's `saveToCassandra` function to insert two new rows into the table kv2 in the keyspace test.

```scala
scala> collection.saveToCassandra("test", "kv", SomeColumns("key", "value"))
scala>
```

You confirm that the two rows have been inserted into the table `test.kv` by querying that table.

```
cqlsh> select * from test.kv;
 key     | value
--------+-------
  cat  |     30
  fox  |     40
...
cqlsh>
```

Voila! Thanks to the *spark-cassandra-connector*, you can now work with Cassandra table data from within a Spark cluster.

You can do a lot with the Cassandra Spark integration. You can combine the computing power of Spark with the storage processing capabilities of Cassandra to do cool things. However, all that is the domain of a different book, so we move on.

Cassandra and the SMACK Stack: A Growing Trend

While big data projects initially started off with a heavy orientation towards batch processing, current trends are towards agile systems that process streaming data to support analytic systems. Organizations are increasingly focusing on real-time or near real-time use cases such as recommendation engines and fraud detection.

SMACK (a system that uses the open source technologies Apache Spark, Apache Mesos, Akka, Apache Cassandra, and Apache Kafka), or variations of such open source technologies, are increasingly becoming crucial players in the analysis and mining of streaming data. Technologies such as SMACK are becoming the platform for constructing applications that make use of "live", streaming data so an organization can effectively extract meaningful information from streaming data sets.

This book is dedicated to an exposition of the concepts and techniques for a Cassandra administrator, and as such, I can't go into the details of all the components of the SMACK stack. However, I'd like to introduce you to this crucial set of technologies since you'll likely to make its acquaintance eventually if you're working with Cassandra.

Here's what the five components of a SMACK stack do:

- *Apache Spark (the processing component)*: Spark is a processing engine for distributed, large-scale data sets and is especially suitable for machine learning algorithms. You use the same interface for SQL queries, machine learning, and stream data processing.

- *Apache Mesos (the container component)*: Mesos is a cluster resource management system that offers resource isolation and sharing to distributed applications. Mesos uses a twin-level scheduling mechanism wherein it offers resources to *frameworks*, which are applications that run on top of Mesos. While Mesos determines the resources that each framework gets, the frameworks determine the resources they accept and which applications they execute on the resources allocated to them.

- *Akka (provides the model)*: Akka is a toolkit and runtime for building highly concurrent and resilient message-driven applications. An "actor" is a universal primitive of concurrent computation that responds to messages it receives. In response to messages, actors can make local decisions, create other actors, send more messages, and so on.

- *Apache Cassandra*: By now, you should know what Cassandra is and what it's good for!

- *Apache Kafka (the broker)*: Kafka is a popular high-volume, low-latency messaging system for handling real-time data feeds. Kafka partitions the data streams and spreads them over a cluster, which also allows for multiple coordinated "consumers" to consume the data, which is generated by "producers."

Summary

This chapter is about showing some interesting ways in which you can set up either test or production environments using different techniques. Docker containers are popular and you can run production Cassandra clusters on Docker. When running a production cluster on Docker, you need to deal with storage and networking strategies, which I haven't discussed here. Docker's documentation is quite good, and you may want to use it to learn more about these and other Docker container topics.

Apache Spark is becoming increasingly popular as a fast processing framework, and I showed how to connect to a Cassandra database from Spark and perform DML operations. Of course, this is only a beginning, and if you wish to delve deeper into Apache Spark, the source documentation is pretty good.

CCM is a cool tool because it helps you spin up clusters in just a few short minutes, without having to learn anything about Cassandra installation procedures and without spending time setting up any prerequisites for running a simple test cluster for exploratory purposes.

Backup, Recovery, and Moving Data

Backing up and restoring data are key functions of a Cassandra administrator. This chapter shows you how to back up your data by taking periodic snapshots of data. Cassandra also enables you to take incremental backups and back up the commit log segments. All these backups come in handy when you need to perform a full restore of the database.

Bulk loading and copying of data is also a common task, and I explain how to load data with the COPY command, as well as with the sstableloader utility.

Backing Up Data

Backing up a Cassandra database involves the backing up of all the SSTable files in the data directory of the database. You can take the backups while the database is running. You can back up the entire database, or just a keyspace, or even a single table.

The backup you make (snapshots) may not be consistent, but that's not a problem at all. The purpose in taking a backup is so you can restore the database if you happen to lose part or all of the data for any reason. When you restore a snapshot, Cassandra utilizes the consistency mechanisms you learned about in earlier chapters to make the data consistent across all nodes in the cluster.

The nodetool snapshot command enables you to back up Cassandra's data. The tool backs up both the schema information and the data in the SSTables. When you need to restore tables from the snapshot, you'll need the data as well as the schema. This way, when you need to recreate some or all of a node's SSTables, you'll have both the data and metadata to create the table(s).

© Sam R. Alapati 2018
S. R. Alapati, *Expert Apache Cassandra Administration*, https://doi.org/10.1007/978-1-4842-3126-5_8

Taking a snapshot means an increase in disk usage since you're storing the same data in two places on the nodes. The nodetool snapshot command works only at a single-node level. You can use a parallel shell utility such as pdsh to run a cluster-wide snapshot by running the nodetool snapshot command.

You can back up all the keyspaces on a node with the following command:

```
$ nodetool snapshot
Requested creating snapshot(s) for [all keyspaces] with snapshot name
[1499019048455] and options {skipFlush=false}
Snapshot directory: 1499019048455
$
```

Cassandra takes the snapshot, which contains the data and the schemas for all tables on this node, and stores the data for each table in a directory corresponding to that table.

```
$ CASSANDRA_HOME/data/data/keyspace_name/table_UID/snapshot_name
```

In this case, it's the following directory:

```
data/data/cycling/cyclist_name-351e8df05f4011e7b2672fabda59ff45/
snapshots/1499019048455
```

You'll find all the files pertaining to the table cyclist_name in this directory:

```
# ls -altr
total 52
-rw-r--r-- 4 root root   92 Jul  2 11:03 mc-3-big-Summary.db
-rw-r--r-- 4 root root   60 Jul  2 11:03 mc-3-big-Index.db
-rw-r--r-- 4 root root   16 Jul  2 11:03 mc-3-big-Filter.db
-rw-r--r-- 4 root root    9 Jul  2 11:03 mc-3-big-Digest.crc32
-rw-r--r-- 4 root root  144 Jul  2 11:03 mc-3-big-Data.db
-rw-r--r-- 4 root root   43 Jul  2 11:03 mc-3-big-CompressionInfo.db
-rw-r--r-- 4 root root   92 Jul  2 11:03 mc-3-big-TOC.txt
-rw-r--r-- 4 root root 4668 Jul  2 11:03 mc-3-big-Statistics.db
-rw-r--r-- 1 root root  856 Jul  2 11:03 schema.cql
-rw-r--r-- 1 root root   31 Jul  2 11:03 manifest.json
drwxr-xr-x 2 root root 4096 Jul  2 11:03 .
drwxr-xr-x 7 root root 4096 Jul  2 11:10 ..
$
```

All files in this directory have a .db extension, denoting that they're all part of the SSTable named CYCLIST_NAME. The schema.cql file contains the complete CQL code to create this table.

You can also back up a single keyspace with the nodetool snapshot command by specifying the keyspace name.

```
$ nodetool snapshot cycling
Requested creating snapshot(s) for [cycling] with snapshot name
[1499020818525] and options {skipFlush=false}
Snapshot directory: 1499020818525
$
```

Managing Snapshots

You can list all snapshots on a node, along with their name and size details, by executing the nodetool listsnapshots command.

```
$ nodetool listsnapshots
Snapshot Details:
Snapshot name Keyspace name      Column family name
True size Size on disk
1499021840269 system_distributed parent_repair_history
23.69 KiB 23.72 KiB
1499021840269 system_distributed repair_history
5.69 KiB  5.72 KiB
1499021840269 system_distributed view_build_status
0 bytes    13 bytes
1499021840269 cycling            cyclist_name
5 KiB      5.87 KiB
1499021840269 test5              kv
4.78 KiB  5.62 KiB
1499021840269 test               my_table
0 bytes    842 bytes
1499021840269 cycling2           cyclist_name
0 bytes    870 bytes
```

```
1499021840269 system_auth          roles
4.95 KiB  4.98 KiB
1499021840269 system_auth          role_members
0 bytes    13 bytes
1499021840269 system_auth          resource_role_permissons_index
0 bytes    13 bytes
1499021840269 system_auth          role_permissions
0 bytes    13 bytes
1499021840269 system_traces        sessions
0 bytes    13 bytes
1499021840269 system_traces        events
0 bytes    13 bytes

Total TrueDiskSpaceUsed: 44.11 KiB
$
```

Snapshots occupy space. They also get dated. This means that you must often clear old snapshot files that you no longer need. You do this by removing the entire snapshot directory. You can script your backups such that you delete the old snapshots before taking a fresh snapshot.

The nodetool clearsnapshot command lets you delete snapshots you've made for one or more keyspaces.

You can delete a single snapshot by running the nodetool clearsnapshot command and specifying the snapshot directory name.

```
$ ./nodetool clearsnapshot 1499020818525
Requested clearing snapshot(s) for [1499020818525]
$
```

You can remove all the snapshot directories on a node by issuing the nodetool clearsnapshot command without a snapshot name.

```
$./nodetool clearsnapshot
Requested clearing snapshot(s) for [all keyspaces]
$
```

Here are some important options you can specify for the `nodetool clearsnapshot` command:

- `-t`: Enables you to specify a snapshot name.

- `keyspace`: Delete snapshots from the keyspace you specify. You can specify multiple keyspaces, each separated by a space.

- `snapshot`: The name of the snapshot you want to delete.

In a production setting, to make a database backup, you take a snapshot, and zip up the files (tar) and store the zip files in network backup locations.

Taking an Automatic Snapshot Before Compacting Data

You can enable the database to take an automatic snapshot before performing a compaction of SSTables (I explain the compaction feature in detail in Chapter 11). You do this by setting the `snapshot_before_compaction` property to true in the `cassandra.yaml` file. The default value is `false`.

Remember that since the database doesn't automatically remove older snapshots, you must be aware of the space consequences of the automatic snapshots.

Performing Incremental Backups

Once you create a system-wide snapshot, you can employ Cassandra's incremental backups feature to back up data that has been updated since the full snapshot was made.

Incremental backups are disabled by default, and you can enable them with the following command:

```
$ nodetool enable backup
```

Alternatively, you can enable incremental backups on a long-term basis by changing the value of the incremental-backups property to `true` in the `cassandra.yaml` file. You can disable incremental backups with the following command:

```
$ nodetool disable backup
```

Incremental backups are automatic. Cassandra creates a hard link to each SSTable that was flushed to disk in the `backups` subdirectory of the `data` folder of the keyspace to which the table belongs. Cassandra doesn't remove the hard links, however, so the administrator must tend to these.

Note   Run the `nodetool statusbackup` command to get the status of a backup.

Restoring Data with Various Restore Methods

To fully restore data, you must, of course, have a complete backup, which consists of the following:

- A snapshot at a point in time

- All incremental backups from the time you took the snapshot

- All commitlog segments since the time you took the last incremental backup

Restoring Data from a Snapshot

To restore a table from a snapshot, make sure you have all the snapshot files for the table, including any incremental backups made after you took the initial snapshot. There are a couple of ways to restore with snapshots, and I explain them here.

In both restore methods, the table schema must be present in the database; the recovery process doesn't recreate the schemas automatically. You can, however, run the `createschema.sql` script found in the snapshot directory to create the schema, should you need it.

Copying Data from the Snapshots Directory

1. Truncate the table you are going to restore. You need to remove the existing data in most cases, since it's possible that accidentally lost data may have older tombstones than the snapshot's data. In situations where you happen to lose a disk and start the database before a restore, the node will have more up-to-date data than the snapshot, and therefore you don't truncate the table.

In this example, the keyspace is *cycling*, and the table name is cyclist_name. Before you truncate the table, go to the data directory for the cyclist keyspace, and you'll see that the data folder for the table cyclist_name has all the normal db files.

```
$CASSANDRA_HOME/data/data/cycling# cd c*
samalapati@ubuntu:/cassandra/apache-cassandra-3.10/data/data/
cycling/cyclist_name-39cd6de060de11e7805be14006afbdda# ls
backups                        mc-1-big-Data.db
mc-1-big-Filter.db   mc-1-big-Statistics.db   mc-1-big-TOC.txt
mc-1-big-CompressionInfo.db   mc-1-big-Digest.crc32
mc-1-big-Index.db    mc-1-big-Summary.db       snapshots
$
```

Note that the snapshots directory for an SSTable will be empty by default until you create a snapshot

2. You next truncate the table cyclist_name.

```
cqlsh:cycling> truncate cyclist_name;
cqlsh:cycling>
```

Note   Make sure all nodes are up before issuing the truncate command. Otherwise you may see an error such as the following:

```
TruncateError: Error during truncate: Cannot achieve
consistency level ONE
```

---

3. You can verify that there's no longer any data in the cyclist_name table.

```
cqlsh:cycling> select * from cyclist_name;
 id | firstname | lastname
----+-----------+----------

(0 rows)
cqlsh:cycling>
```

4. Check the data directory for the cycling.cyclist_name table now; all the .db files disappeared from this directory when you truncated the table. Only the empty shell of the table remains, but the database has wiped out all its contents from disk for good.

You took a snapshot of the *cycling* keyspace before you truncated the cyclist_name table. You therefore expect to have the snapshot for the cycling_name table in the appropriate directory, which in general is:

```
data_directory/keyspace_name/table_name-UUID/snapshost/
snapshot_name
```

In my case, it's the following directory, and it has all the files I need under the snapshot directory for the table cyclist_name:

```
data/data/cycling/cyclist_name-39cd6de060de11e7805be14006a
fbdda/snapshots/1499189424022# ls -altr
total 52
-rw-r--r-- 2 root root   16 Jul  4 10:30 mc-1-big-Filter.db
-rw-r--r-- 2 root root   92 Jul  4 10:30 mc-1-big-TOC.txt
-rw-r--r-- 2 root root   92 Jul  4 10:30 mc-1-big-Summary.db
-rw-r--r-- 2 root root 4668 Jul  4 10:30 mc-1-big-
Statistics.db
-rw-r--r-- 2 root root   60 Jul  4 10:30 mc-1-big-Index.db
-rw-r--r-- 2 root root   10 Jul  4 10:30 mc-1-big-Digest.
crc32
-rw-r--r-- 2 root root  155 Jul  4 10:30 mc-1-big-Data.db
-rw-r--r-- 2 root root   43 Jul  4 10:30 mc-1-big-
CompressionInfo.db
-rw-r--r-- 1 root root  856 Jul  4 10:30 schema.cql
-rw-r--r-- 1 root root   31 Jul  4 10:30 manifest.json
drwxr-xr-x 3 root root 4096 Jul  4 10:30 ..
drwxr-xr-x 2 root root 4096 Jul  4 10:30 .
/data/data/cycling/cyc
```

5. Copy all the files under the snapshot directory to the data directory of the table cyclist_name.

    ```
    root@ubuntu:/cassandra/apache-cassandra-3.10/data/
    data/cycling/cyclist_name-39cd6de060de11e7805be14006a
    fbdda# cp /cassandra/apache-cassandra-3.10/data/data/
    cycling/cyclist_name-39cd6de060de11e7805be14006afbdda/
    snapshots/1499189424022/* .
    root@ubuntu:/cassandra/apache-cassandra-3.10/data/data/
    cycling/cyclist_name-39cd6de060de11e7805be14006afbdda# ls
    1499189424022   manifest.json              mc-1-big-Data.db
    mc-1-big-Filter.db  mc-1-big-Statistics.db  mc-1-big-TOC.txt
      snapshots
    backups         mc-1-big-CompressionInfo.db  mc-1-big-
    Digest.crc32  mc-1-big-Index.db    mc-1-big-Summary.db
    schema.cql
    root@ubuntu:/cassandra/apache-cassandra-3.10/data/data/
    cycling/cyclist_name-39cd6de060de11e7805be14006afbdda#
    ```

    Run the nodetool refresh command so Cassandra knows that the data files are restored now.

    ```
    $ nodetool refresh cycling cyclist_name
    $
    ```

    The Cassandra log file will show how many SSTables were refreshed, thus offering a safe way to check the refresh process.

6. Query the table cyclist_name to verify that the restore was successful.

```
cqlsh:cycling> select * from cyclist_name;
 id                                   | firstname | lastname
--------------------------------------+-----------+-----------
 fb372533-eb95-4bb4-8685-6ef61e994caa |   Michael |  MATTHEWS
 220844bf-4860-49d6-9a4b-6b5d3a79cbfb |     Paolo | TIRALONGO
 6ab09bec-e68e-48d9-a5f8-97e6fb4c9b47 |    Steven | KRUIKSWIJK

(3 rows)
cqlsh:cycling>
```

Note   You can set the `auto_snapshot` property to automatically back up (have a snapshot taken) whenever you truncate or drop a table. By default, the `auto_snapshot` property is enabled.

Hey, what happened? There were six rows originally in this table, but after the restore, there are just three rows! Well, there are two nodes in the cluster, and each node owns roughly half the data. The `nodetool snapshot` command did create a separate snapshot directory in both nodes. The snapshot on any one node won't have all the data. If, for example, your replication factor is 3 and you have four nodes in your cluster, each node will have approximately 0.75 of each replica.

In this example, there are two nodes in the test cluster. However, you performed a restore on only one node. You must therefore restore the `.db` files on the second node, as you did earlier.

```
# cp /cassandra/apache-cassandra-3.10/data/data/cycling/cyclist_name-39cd6de
060de11e7805be14006afbdda/snapshots/truncated-1499189595147-cyclist_name/* .
# ./nodetool refresh cycling cyclist_name
```

Once you do this, query the `cyclist_name` table again:

```
cqlsh:cycling> select * from cyclist_name;
 id                                   | firstname | lastname
--------------------------------------+-----------+------------------
 e7ae5cf3-d358-4d99-b900-85902fda9bb0 |      Alex |            FRAME
 fb372533-eb95-4bb4-8685-6ef61e994caa |   Michael |         MATTHEWS
 5b6962dd-3f90-4c93-8f61-eabfa4a803e2 |  Marianne |              VOS
 220844bf-4860-49d6-9a4b-6b5d3a79cbfb |     Paolo |        TIRALONGO
 6ab09bec-e68e-48d9-a5f8-97e6fb4c9b47 |    Steven |       KRUIKSWIJK
 e7cd5752-bc0d-4157-a80f-7523add8dbcd |      Anna | VAN DER BREGGEN
(6 rows)
cqlsh:cycling>
```

Tip   If a table doesn't exist before performing a restore, create the table with the help of the `creatschema.cql` file in the `snapshot` directory for the table.

Running a Repair After Restoring from a Snapshot

Cassandra's repair mechanism ensures that all nodes in a cluster have the most recent data. If you haven't taken the snapshots for different nodes at the same time, you can read stale data (if using the consistency level `ONE`),

This is the reason why you should run the `nodetool repair` command after restoring from snapshots. The repair makes sure that all replicas in the cluster have the latest data. The repair could run for several hours, since it's a heavy resource user and requires both RAM and CPU resources to do its job. As part of its work, the repair job generates Merkel trees for the data and compares them with the Merkel trees from the other nodes. It also streams missing and/or outdated data. While the repair is running, you may sometimes read stale data if all the snapshots don't contain the same data.

Using sstableloader to Restore Snapshots

Let's say you have a bunch of tables you want to restore from snapshots. You can use the sstableloader utility to restore the snapshots.

The Node Restart Method Procedure

The node restart method of restoring data from snapshots involves shutting down all the nodes and then starting them back up after you restore the snapshot data. You can use this method for all nodes in a cluster at once or just for a single node.

The following are the steps in restoring data from snapshots that involves a restart of the nodes.

1. Shut down all the nodes of the cluster.

2. Run the `nodetool drain` command to make sure you're not risking a data loss.

3. Remove all files from the `commitlog` directory.

```
samalapati@ubuntu:/cassandra/apache-cassandra-3.10/data/
commitlog# ls
CommitLog-6-1499187293466.log  CommitLog-6-1499187293467.log
samalapati@ubuntu:/cassandra/apache-cassandra-3.10/data/
commitlog# rm *
```

Remove all the .db files under each table's directory. You must do this for all keyspaces you're restoring. Make sure you leave the snapshots and backups directories alone!

When you're done, there should be no .db files in this directory.

```
$ ls
backups   snapshots
$
```

4. Copy all the files in the latest snapshot directory into this directory:

```
$ cp /cassandra/apache-cassandra-3.10/data/data/
cycling/cyclist_name-4d1743b060ef11e7805be14006afbdda/
snapshots/1499196752095/* .
$
```

5. At this point, Cassandra restores the table data across the cluster. To be sure that the data is consistent, run the nodetool repair command.

```
$ sudo odetool repair
[2017-07-04 13:19:20,492] Replication factor is 1. No
repair is needed for keyspace 'cycling'
 [2017-07-04 13:19:26,018] Repair completed successfully
[2017-07-04 13:19:26,064] Repair command #1 finished in
5 seconds
```

Commitlog Archiving and Point-in-Time Recovery

Cassandra enables you to configure commit log archiving and point-in-time recovery. Cassandra archives the commit log under the following conditions:

- When a node starts up

- When the database writes the commit log to disk

- Any specified point in time

You can configure the archiving of the commit log through the `commitlog_archiving.properties` configuration file. You can set several properties pertaining to the archiving and restoring of commit log segments in this file, as I explain in the following sections.

Manually Archiving the Commit Log

You can archive a commit log segment with the `archive_command` command. Here's the syntax of the `archive_command` command:

```
archive_command=/bin/ln %path /backup/%name
```

In this command, the two key parameters are

- `path`: Fully qualified path to the commit log segment you wish to archive

- `name`: The name of the commit log

Restoring a Commit Log

The archives you make of the commit log segments enable you to recover the database to a point in time. You can restore an archived commit log with the command `restore_command`. Here's the syntax of the command:

```
restore_command=cp -f %from %to
```

In this command, the key parameters are

- `From`: The path to the archived commitlog segment

- `To`: Name of the commit log directory

Setting the Restore Directory Location

You can configure the `restore_directories` parameter to set the location of the directory where you want to store the archived commit log. When you run the `restore_command` command, the database looks here for the archived commitlog segments it needs to restore.

Setting the Restore Timestamp

You can tell the database how far it should restore the commit log segments, by specifying a timestamp with the `restore_point_in_time` parameter. Here's an example:

```
restore_point_in_time=2013:12:11 17:00:00
```

Cassandra will restore the commit log segments that it created prior to and up to the timestamp you specify.

Loading Bulk Data into Cassandra

There are two basic tools for moving bulk data to and from a Cassandra cluster: the CQL COPY command and the sstableloader tool.

- The `COPY` command can read CSV data to a Cassandra table and write CSV data from Cassandra to a file system. The `COPY` command works very similarly to the tools used by relational databases for exporting and importing data.

- The sstableloader utility helps you bulk load external data into a Cassandra cluster.

Using the COPY Command to Import and Export Data

The `COPY` command enables you to copy data to and from a Cassandra database. The source and target files are CSV (comma-separated values) or delimited text files. In the following sections, I review how you can use the COPY command to import and export data.

Copying Data from a Cassandra Table

Use the COPY TO command to copy data from a Cassandra table into a CSV file. The following is the syntax of the COPY TO command:

```
COPY table_name [( column_list )]
TO 'file_name'[, 'file2_name', ...] | STDOUT
[WITH option = 'value' [AND ...]]
```

The COPY TO command separates the fields in the CSV file with a delimiter that you can specify. By default, this command exports all fields, and you can specify a column list to export only part of a table's rows.

Here's an example that shows how to export data from the table CYCLNG.CYCLIST_NAME:

```
cqlsh> COPY cycling.cyclist_name TO 'cyclist.csv' WITH HEADER = TRUE;Using
1 child processes
Starting copy of cycling.cyclist_name with columns [id, firstname,
lastname].
Processed: 6 rows; Rate:       8 rows/s; Avg. rate:       7 rows/s
6 rows exported to 1 files in 0.849 seconds.
cqlsh>
```

Cassandra places the CSV file one directory above the directory from where you executed the COPY TO command. If you open the CSV file, you can see that it has a comma-delimited list of all the rows in the source table.

```
cat cyclist_lastname.csv
id,lastname
fb372533-eb95-4bb4-8685-6ef61e994caa,MATTHEWS
220844bf-4860-49d6-9a4b-6b5d3a79cbfb,TIRALONGO
e7cd5752-bc0d-4157-a80f-7523add8dbcd,VAN DER BREGGEN
6ab09bec-e68e-48d9-a5f8-97e6fb4c9b47,KRUIKSWIJK
e7ae5cf3-d358-4d99-b900-85902fda9bb0,FRAME
5b6962dd-3f90-4c93-8f61-eabfa4a803e2,VOS
```

Copying Data to a Cassandra Table

The COPY command helps move small data sets (a data set that containing less than 2 million rows) into a Cassandra table. You must use the Cassandra bulk loader to move larger data sets.

Specify COPY FROM to copy data from a CSV file into a Cassandra table. The following is the general syntax for the COPY FROM command:

```
COPY table_name [( column_list )]
FROM 'file_name'[, 'file2_name', ...] | STDIN
[WITH option = 'value' [AND ...]]
```

The COPY FROM command updates existing records and verifies the primary key. The following are the key things you need to remember when running the COPY FROM command:

- Each row in the CSV file must contain the same fields.

- The CSV file can have fewer fields (columns) than the Cassandra table.

- The CSV file can't have more fields than the Cassandra table.

- Each row must have a value for the primary key value.

- The database sets all missing or empty fields to null.

For an example of the COPY FROM command, use the same table, CYCLIST_NAME, that you used for the COPY TO command example. Since the COPY FROM command expects you to have the table present, truncate the data from the table, thus making it ready to import the data from the CSV file that you generated earlier.

```
cqlsh> COPY cycling.cyclist_name FROM 'cyclist.csv' WITH HEADER = TRUE;
Using 1 child processes

Starting copy of cycling.cyclist_name with columns [id, firstname,
lastname].
Processed: 6 rows; Rate:     11 rows/s; Avg. rate:     15 rows/s
6 rows imported from 1 files in 0.389 seconds (0 skipped).
cqlsh>
```

Running sstableloader to Perform Bulk Loading

Cassandra's bulk loader tool, named sstableloader, lets you load large amounts of data into Cassandra tables. You can also use this tool to load data from an SSTable to a different cluster. Finally, you can use this tool to restore snapshots, which I explained earlier in this chapter.

Unlike in the case of the COPY command, the target table can have data.

Loading External Data with sstableloader

The sstableloader utility enables you to load external data into your cluster. Before you can load external data with the sstableloader utility, you must first generate the SSTables, into which you put the external data.

To generate the SSTables, you must use the SSTableWriter API. SSTableWriter creates raw Cassandra data files that you can bulk load into a cluster.

Note   You can run more than one instance of the sstableloader utility to parallelly load external data into your cluster.

Since the sstableloader utility is resource intensive, you should run it from servers that aren't acting as a Cassandra node.

The following is a summary of the steps you must follow to load data through the sstableloader utility.

1. Create the raw Cassandra data files that you'll be using to bulk load the data into the cluster. This involves writing some Java code that uses the SSTableWriter API.

2. SSTableWriter generates SSTables in the directory you specify. Go to the location where the code has stored the SSTables, as shown here:

   ```
   $ cd /var/lib/cassanddra/data/Keyspace1/Standrad2
   ```

3. You can view the keyspace's contents by executing the ls command.

   ```
   $ ls
   Keyspace1-Standard1-jb-60-CRC.db
   Keyspace1-Standard1-jb-60-Data.db
   ...
   Keyspace1-Standard1-jb-60-TOC.txt
   ```

4. Bulk load the files by specifying a path to Keyspace1/Standard1/ in the cluster where you wish to load the SSTables.

```
$ sstableloader -d 192.168.159.129 /var/lib/Cassandra/
data/Keyspace1/Standard1/
```

Loading SSTable Data to a Different Cluster

You can run the sstableloader utility to import existing SSTables from a different cluster. Before you perform the import, run the nodetool flush command in each of the source cluster's nodes to ensure that the database flushes the memtables to the SSTables that you're about to import to your cluster.

To run the sstableloader command to import the data to a target cluster, follow the steps I described in the previous section. Everything works the same way, except that you don't need to generate the SSTables since the source table already has them. You are importing from an SSTable to another SSTable in a different cluster, so no raw data is involved in this import.

Note   You can use Apache Sqoop to migrate external data to Cassandra if you're using DataStax Enterprise, which comes loaded with everything you need to run Sqoop.

Summary

Since you need snapshots when you need to perform a recovery, it's important to carefully manage snapshots. Cassandra doesn't manage snapshots for you and therefore you must develop an effective way to store snapshots and remove older snapshots so they don't occupy a lot of storage space.

Incremental backups are turned on by default, and you must ensure that they remain in effect since they reduce the need to take full snapshots of SSTables.

You'll need all snapshots, incremental backups, and backups of the commit log segments during a recovery. Therefore, it's important to know that you can access all of these readily so you can minimize the recovery time. To be sure that you can perform a recovery correctly, it's important to perform dry runs of a recovery session. You don't want to learn database recovery when you're actually performing one!

PART III

Maintaining, Monitoring, Tuning, and Securing Apache Cassandra

Maintaining Cassandra

This chapter is about common Cassandra management tasks and node management.

Common cluster management tasks include repairing and cleaning data, rebuilding indexes, handling data corruption, and flushing and draining data.

Node management includes adding and removing nodes, replacing and moving nodes, removing dead nodes, and assassinating nodes. Since nodetool commands are key to performing all node management tasks, I first review the key nodetool node management-related commands.

The chapter shows how to perform management tasks such as adding and decommissioning datacenters.

The chapter also shows you how to switch snitches and how to monitor and manage gossip.

Common Cluster Maintenance Tasks

As an administrator, you'll perform various tasks to help Cassandra perform well. In the following sections, I describe some of the most common database maintenance tasks, such as repairing data, rebuilding indexes, and cleaning up data.

Repairing Data (nodetool repair)

I explained Cassandra's repair mechanisms at length in Chapter 5. Over time, a node may become out of sync with other nodes due to Cassandra's tuneable consistency philosophy. If a node remains down for a while, it may miss changes performed on the other nodes, especially if it stays down longer than the time for which the other nodes store the hints.

© Sam R. Alapati 2018
S. R. Alapati, *Expert Apache Cassandra Administration*, https://doi.org/10.1007/978-1-4842-3126-5_9

In addition, specifying a write consistency level below the ALL level means that the database declares a write as successful even if one of more nodes don't acknowledge the writes. The principle of tunable consistency could result in nodes with different versions of the same data. While the database automatically performs some types of repair such as read repair, one of the key operations an administrator performs is anti-entropy repair, which involves manually repairing data with the `nodetool repair` command.

Tip   Run the `nodetool stats` command while performing a repair to monitor the progress of the repair.

When to Perform a Repair

The frequency of repairs has a lot to do with the read and write consistency levels that you configure. If you chose a consistency level that's slow to provide consistency, then you should schedule repairs more frequently.

Frequent repairs have a cost in terms of overhead, so use a subrange repair wherever you can, and schedule the repairs for different keyspaces (or tables) for separate times.

When Repairs Are Unavoidable

Although the repairs you perform with the `nodetool repair` command are optional, there are several operations that require that you perform a full repair after completing the operations. These operations include changing a cluster's snitch and changing the replication level of keyspaces.

Running the nodetool repair Command

You can run the `nodetool repair` command for the entire database or for a specific keyspace or table, as shown in the following examples:

```
$ nodetool repair
$ nodetool repair cycling
$ nodetool repair cycling cycle
```

You can also run the nodetool repair command just for the local datacenter with the -local (or –in-local-dc) option. You can also specify a datacenter with the -dc (or –in-dc) option.

Cassandra performs a full repair when you specify the -full option when you run the nodetool repair command. Since a full repair can be time consuming and resource intensive, you can ask it to perform an incremental repair, wherein the database only repairs previously unrepaired data. The default is incremental repair in Cassandra 3.x. Here's an excerpt from the log file that shows how the repair command works:

```
INFO  [Thread-8] 2017-07-16 09:22:23,564 RepairRunnable.java:136 - Starting
repair command #1 (f30498c0-6a42-11e7-9731-6f6038c9cb1c), repairing
keyspace system_traces with repair options (parallelism: parallel, primary
range: false, incremental: true, job threads: 1, ColumnFamilies: [],
dataCenters: [], hosts: [], # of ranges: 512, pull repair: false)
...
INFO  [Repair#1:1] 2017-07-16 09:22:27,833 RepairJob.java:172 - [repair
#f35db540-6a42-11e7-9731-6f6038c9cb1c] Requesting merkle trees for sessions
(to [/192.168.159.129, ubuntu/192.168.159.130])
...
INFO  [CompactionExecutor:9] 2017-07-16 09:22:28,547 CompactionManager.
java:694 - [repair #f30498c0-6a42-11e7-9731-6f6038c9cb1c] Completed
anticompaction successfully
INFO  [CompactionExecutor:10] 2017-07-16 09:22:28,759 RepairRunnable.
java:340 - Repair command #1 finished in 5 seconds
$
```

Rebuilding Data by Getting It from Other Nodes (nodetool rebuild)

You run the nodetool rebuild command to rebuild a token range. The command rebuilds streams from a single source replica but operates simultaneously on multiple nodes.

The nodetool rebuild command is useful when you're adding a datacenter to your cluster. You can rebuild a single keyspace at a time or specify multiple keyspaces, each separated by a comma. Here's an example:

```
$ nodetool rebuild –keyspace cyclists, motorists
```

Remember:

- By default, this command choses an arbitrary datacenter in your cluster. You can specify a datacenter from which the database should select the sources for the data it needs to stream through the source-dc-name option. If you don't provide the name of any datacenter, the command may seem to work, but it won't do anything.

- You can optionally specify the parameter tokens to provide a list of single tokens or a range of tokens (start_token, end_token).

Cleaning Up Unnecessary Keyspaces and Partition Keys (nodetool cleanup)

When you add a new node to a cluster, some nodes may lose part of their partition range to the new node. However, Cassandra doesn't remove the unnecessary keyspaces and partition keys from the current nodes when this happens.

Similarly, if you decrease the replication factor for a datacenter, some nodes will end up storing data that isn't being used by the database as replicas.

In both cases, when you add nodes or when you reduce the replication factor, Cassandra's compaction processes will eventually get rid of the discarded and unnecessary data. You don't have to manually clean up after every such change in a cluster. However, if for any reason you wish to reclaim disk space faster, you can run the nodetool cleanup command.

Note   If the node uses a counter column in some tables, once you run the nodetool cleanup command, the database assigns new counter IDs to that node.

Here's an example that shows how to run the nodetool cleanup command:

```
$ nodetool cleanup
```

Since you didn't specify a keyspace, this command will clean up all keyspaces that are no longer part of this node. Optionally, you can specify a keyspace, as well as a list of tables for cleaning up. You run the nodetool cleanup command as follows to clean up a specific keyspace:

```
$ nodetool cleanup cycling
```

Rebuilding Indexes

You can rebuild secondary indexes by running the nodetool rebuild_index command for a table. In the command, you must specify the names of the keyspace, table, and one or more index names separated by a space.

The nodetool rebuild_index command performs a full rebuild of the index for a table. This command requires three arguments: keyspace, cf (column family, which is a synonym for a table), and the name of the index. Here's an example:

```
$ nodetool rebuild_index cycling cycle test_idx;
```

Refreshing Size Estimates of Tables

When you insert large amounts of data into a table or when you truncate a table's data, the table's size estimates become stale. Run the nodetool refreshsizeestimates command to freshen the table size estimates. The command refreshes the system.size_estimates table.

Key Nodetool Maintenance Commands

In this chapter, among other things, I show you how to add, move, and remove nodes in your cluster. When performing most of the tasks pertaining to the adding, moving, and removing of nodes and datacenters in your cluster, you'll be running a set of key nodetool commands. For example, you'll be running nodetool commands after decommissioning nodes. I explain the common node maintenance-related nodetool commands in this section.

Decommissioning Nodes (nodetool decommission)

You run the nodetool decommission command to make a node send all its data to the next node in the ring. Decommissioning a node is the opposite of bootstrapping a node. Here's an example showing how to decommission a node.

1. First, check the status of the nodes to make sure that all (two in this case) the nodes are up.

```
$ /nodetool status
Datacenter: datacenter1
=========================
Status=Up/Down
|/ State=Normal/Leaving/Joining/Moving
--  Address           Load       Tokens       Owns (effective)
Host ID                                Rack
UN  192.168.159.129  338.19 KiB  256              49.0%
0dbb9e0e-867e-4179-b6b6-631d38dd68f9  rack1
UN  192.168.159.130  339.49 KiB  256              51.0%
a1085901-738c-4bbd-b050-007f62da893d  rack1
$
```

2. Then run the nodetool decommission command on the node
    with the IP 192.168.159.129.

    ```
    $ nodetool decommission
    ```

3. Check the status of the cluster and notice that the
    decommissioned node now shows its status as UL (UP LEAVING),
    indicating that it's transitioning out of the cluster.

    ```
    UN  192.168.159.129  338.19 KiB  256              49.0%
    0dbb9e0e-867e-4179-b6b6-631d38dd68f9  rack1
    UL  192.168.159.130  339.49 KiB  256              51.0%
    a1085901-738c-4bbd-b050-007f62da893d  rack1
    $
    ```

    The system.log file shows the following output after the running
    of the nodetool decommission command:

    ```
    CP Connection(4)-127.0.0.1] 2017-07-13 10:48:38,266
    StorageService.java:1435 - LEAVING: sleeping 30000 ms for
    batch processing and pending range setup
    INFO  [RMI TCP Connection(4)-127.0.0.1] 2017-07-13
    10:49:08,776 StorageService.java:1435 - LEAVING: replaying
    batch log and streaming data to other nodes
    ```

```
INFO  [RMI TCP Connection(4)-127.0.0.1] 2017-07-13
10:49:08,902 StreamResultFuture.java:90 - [Stream
#9254bbf0-67f3-11e7-9d74-6bc8e6fb7ba6] Executing streaming
plan for Unbootstrap
INFO  [RMI TCP Connection(4)-127.0.0.1] 2017-07-13
10:49:08,904 StorageService.java:1435 - LEAVING: streaming
hints to other nodes
INFO  [HintsDispatcher:2] 2017-07-13 10:49:08,963
HintsDispatchExecutor.java:152 - Transferring all hints to
0dbb9e0e-867e-4179-b6b6-631d38dd68f9
...
INFO  [RMI TCP Connection(4)-127.0.0.1] 2017-07-13
10:49:09,413 StorageService.java:3938 - Announcing that I
have left the ring for 30000ms
INFO  [RMI TCP Connection(4)-127.0.0.1] 2017-07-13
10:49:39,428 Server.java:176 - Stop listening for CQL
clients
WARN  [RMI TCP Connection(4)-127.0.0.1] 2017-07-13
10:49:39,429 Gossiper.java:1514 - No local state, state is
in silent shutdown, or node hasn't joined, not announcing
shutdown
...
$
```

4. Check the status of the cluster. Sure enough, the decommissioned
    node has left the cluster.

```
$ /nodetool status
Datacenter: datacenter1
========================
Status=Up/Down
|/ State=Normal/Leaving/Joining/Moving
--  Address          Load       Tokens       Owns (effective)
Host ID                          Rack
UN  192.168.159.129  328.53 KiB  256          100.0%
0dbb9e0e-867e-4179-b6b6-631d38dd68f9  rack1
$
```

5. Next, if you try to add the decommissioned node back to the cluster by starting it back up, the attempt fails, as the following output from the `cassandra` command shows:

```
$ cassandra ...
, 9079045138376597795, 9105248878993821997,
9166471019946737685, 9167277031713212107,
9167870256226113023]
...
This node was decommissioned and will not rejoin the
ring unless cassandra.override_decommission=true has
been set, or all existing data is removed and the node
is bootstrapped again
Fatal configuration error; unable to start server.  See
log for stacktrace.
ERROR [main] 2017-07-13 10:58:29,816 CassandraDaemon.
java:752 - Fatal configuration error
...
$
```

The node that you recommissioned is unable to join the ring when you restart the node. As you can see, Cassandra helpfully offers two ways to enable this node to rejoin the ring:

- Remove all data from the decommissioned node and bootstrap it again.

- Set the `cassandra.override_decommission=true` option.

In the following sections, I show how to use both methods to rejoin the decommissioned node to a cluster.

Remove All Data and Restart the Node

The straightforward way to add a decommissioned node back to a cluster is to simply remove all the old data from the node, which is stored under the following directories under /data:

- `commitlog`

- `data`

- `saved_caches`

When you decommission the node, Cassandra redistributes the tokens. Once you remove all the data from the decommissioned node, Cassandra allows the node to rejoin the node and allocates the node its share of the tokens again.

Once you remove all the data, restart the decommissioned node.

```
$ cassandra
...
INFO  [main] 2017-07-13 11:13:47,122 ColumnFamilyStore.java:406 -
Initializing system.hints
INFO  [main] 2017-07-13 11:13:47,156 ColumnFamilyStore.java:406 -
Initializing system.batchlog
INFO  [GossipStage:1] 2017-07-13 11:13:50,736 Gossiper.java:1056 - Node
/192.168.159.129 is now part of the cluster
INFO  [main] 2017-07-13 11:13:51,939 StorageService.java:1435 - JOINING:
waiting for ring information
INFO  [InternalResponseStage:1] 2017-07-13 11:13:52,829 ColumnFamilyStore.
java:406 - Initializing cycling.cycle
INFO  [main] 2017-07-13 11:14:24,036 StorageService.java:1435 - JOINING:
Starting to bootstrap...
INFO  [main] 2017-07-13 11:14:25,128 StorageService.java:1435 - JOINING:
Finish joining ring
...
INFO  [main] 2017-07-13 11:14:25,193 SecondaryIndexManager.java:508 -
Executing pre-join post-bootstrap tasks for: CFS(Keyspace='cycling',
ColumnFamily='cycle')
INFO  [main] 2017-07-13 11:14:33,359 CassandraDaemon.java:725 - No gossip
backlog; proceeding
...
```

The decommissioned node joins the ring now. Cassandra has already distributed the data in the cluster. You confirm that the node is back in the cluster by running the nodetool status command.

```
$ nodetool status
Datacenter: datacenter1
========================
Status=Up/Down
|/ State=Normal/Leaving/Joining/Moving
```

```
--   Address              Load          Tokens          Owns (effective)
Host ID                                  Rack
UN   192.168.159.129   350.33 KiB   256              48.5%
0dbb9e0e-867e-4179-b6b6-631d38dd68f9   rack1
UN   192.168.159.130   158.45 KiB   256              51.5%
9ee09b72-a6db-4de4-9aed-de32c5e7c344   rack1
$
```

Setting the cassandra.override_decommission=true Option

Alternatively, you can just set the following option in the `cassandra-env.sh` file of the node you've decommissioned and restart the node. Once the node restarts, it'll join the cluster.

```
JVM_OPTS="$JVM_OPTS -D cassandra.override_decommission=true
```

Assassinating a Node (nodetool assassinate)

Let's say you have three nodes in your cluster and you want to remove one of the nodes. You decommission the node, but when you run the `nodetool describe cluster` command, you see the node in an UNREACHABLE state, as shown here:

```
$ nodetool describecluster
Cluster Information:
        Name: Test Cluster
        Snitch: org.apache.cassandra.locator.DynamicEndpointSnitch
        Partitioner: org.apache.cassandra.dht.Murmur3Partitioner
        Schema versions:
                57af9326-0783-3ea2-93ec-7706e8cad5e7: [192.168.159.129]

                UNREACHABLE: [192.168.159.130]
$
```

The `nodetool status` command shows the unreachable node with a DN status, indicating that the node is down.

```
$ nodetool status
Datacenter: DC1
```

```
================
Status=Up/Down
|/ State=Normal/Leaving/Joining/Moving
--  Address          Load        Tokens        Owns (effective)
Host ID                                Rack
DN  192.168.159.130  ?           256           49.7%
7d9e3ac3-ffdc-4cee-be82-4b0c6e614aa1  r1
Datacenter: datacenter1
=======================
Status=Up/Down
|/ State=Normal/Leaving/Joining/Moving
--  Address          Load        Tokens        Owns (effective)
Host ID                                Rack
UN  192.168.159.129  383.92 KiB  256            50.3%
0dbb9e0e-867e-4179-b6b6-631d38dd68f9  rack1
$
```

In cases such as this, where you're unable to take a node out of a cluster with the nodetool decommission command, run the nodetool assassinate command to get rid of the node.

```
$ nodetool assassinate
```

Check the cluster status with the nodetool status command to ensure that the node you've assassinated (192.168.159.130) is gone from the cluster.

```
$ nodetool status
Datacenter: datacenter1
=======================
Status=Up/Down
|/ State=Normal/Leaving/Joining/Moving
--  Address          Load        Tokens        Owns (effective)
Host ID                                Rack
UN  192.168.159.129  359.04 KiB  256            100.0%
0dbb9e0e-867e-4179-b6b6-631d38dd68f9  rack1
$
```

Run the nodetool assassinate command whenever you fail to remove a node with the nodetool removenode command. The nodetool assassinate command removes a dead node without replicating the data.

The nodetool assassinate command removes the node where you run the command. You can specify a remote node with the -h option.

```
$ nodetool assassinate -h 192.168.159.129
```

Node Management

Node management is an integral part of a Cassandra administrator's task list. Managing nodes involves adding, moving, and replacing nodes. It also involves managing slow or failed nodes and cleaning up data after range movements.

In the following sections, I explain how to perform key node management tasks.

Keeping a Node from Joining the Cluster

At times, you don't want a node to join the cluster. That is, you want the node to be up and running, but without being part of the ring. You can do this by utilizing the Cassandra command line utility's -D option. The -D option enables you to specify startup parameters in the cassandra-env.sh file and at the command line. In this case, add the -Dcassandra.join.ring=false option to the cassandra-env.sh file.

The following are the steps you must follow to keep a node from joining a cluster.

1. Stop the node you wish to take out of the ring.

2. Edit the cassandra-env.sh file and add the following line:

   ```
   JVM_OPTS="$JVM_OPTS -Dcassandra.join_ring=false"
   ```

3. Start up the node.

   ```
   $ cassandra -R
   ...
   INFO  [main] 2017-07-15 12:31:36,878 CassandraDaemon.
   java:489 - JVM Arguments: [-...
   Djava.library.path=./../lib/sigar-bin,
   -Dcassandra.join_ring=false, -
   ```

```
...
INFO  [GossipStage:1] 2017-07-15 12:31:44,587 Gossiper
.java:1056 - Node /192.168.159.129 is now part of the
cluster
INFO  [GossipStage:1] 2017-07-15 12:31:44,589 Gossiper.
java:1056 - Node /192.168.159.130 is now part of the
cluster
INFO  [main] 2017-07-15 12:31:45,757 StorageService.
java:695 - Not joining ring as requested. Use JMX
(StorageService->joinRing()) to initiate ring joining
INFO  [main] 2017-07-15 12:31:45,758 CassandraDaemon.
java:694 - Waiting for gossip to settle before accepting
client requests...
INFO  [GossipStage:1] 2017-07-15 12:31:46,429
StorageService.java:2248 - Node /192.168.159.129 state
jump to NORMAL
INFO  [main] 2017-07-15 12:31:53,763 CassandraDaemon.
java:725 - No gossip backlog; proceeding
$
```

As you can see from the Cassandra utility's output, the node (192.168.159.130) has
started up fine and is part of the cluster. However, since you asked Cassandra not to
join this node to the cluster, it remains up but out of the cluster. You can confirm this by
running the nodetool status command.

```
$ nodetool status
Datacenter: datacenter1
========================
Status=Up/Down
|/ State=Normal/Leaving/Joining/Moving
--  Address           Load        Tokens        Owns (effective)
Host ID                             Rack
UN  192.168.159.129  296.03 KiB   256             50.3%
0dbb9e0e-867e-4179-b6b6-631d38dd68f9  rack1
DN  192.168.159.130  316.55 KiB   256             49.7%
7d9e3ac3-ffdc-4cee-be82-4b0c6e614aa1  rack1
$
```

The node 192.168.159.130 is showing a status of DN (DOWN NORMAL). The node is up but is no longer a part of this cluster.

Subsequently, you can run the `nodetool join` command to add this node back to the ring, as shown here:

```
$ nodetool join
```

Adding a Node to a Data Center

Adding a new node involves installing and starting up the new node, and cleaning up the existing nodes to remove keys that no longer belong to those nodes. Follow these steps to add a node to a running cluster.

1. Install the Cassandra binaries on the new node.

2. Configure the following properties in the `cassandra.yaml` file:

    - `cluster_name`: The cluster name.

    - `endpoint_snitch`: GossipingPropertyFileSnitch (or any other snitch you wish to use).

    - `num_tokens`: 256 (this sets the number of vnodes, and you can set it to the same number as the tokens you configured for the rest of the nodes)

    - `allocate_tokens_for_local_replication_factor`: Enables you to specify the replication factor of the keyspaces.

    - `seed_provider`: Specify one of the existing seed nodes in the cluster.

3. Ensure that you add any non-default settings you configured in the `cassandra.yaml`, `cassandra-rackdc.properties`, and `cassandra-topology.properties` files. All these configuration files on the new node should use the same non-default settings as the rest of the nodes.

4. Start the new node.

5. Verify the status of the nodes with the `nodetool status` command and ensure that all nodes show an UN (Up Normal) status.

6. Run the `nodetool cleanup` command on all the existing nodes. You must wait until the command completes on one node before proceeding to the other nodes.

This scenario is for adding a new node to a cluster. If you're replacing an existing node that died for some reason, you need to let Cassandra know that the new node you're adding will be replacing the dead node. You do this by specifying the `replace_address` option.

Stop the node you wish to replace, before performing one of the following steps.

- For package installations, add the `replace_address` option to the `cassandra-env.sh` file before starting the new node (see Step 4).

  ```
  JVM_OPTS = "$JVM_OPTS -Dcassandra.replace_
  address=192.168.159.130
  ```

- For a tarball installation, as in my case, start the database with the `cassandra.replace_address` option.

You can specify the `listen_address` or the `broadcast_address` of the dead node as the value for the `cassandra.replace_address` option. You must ensure that the node you're replacing the other node with is either brand now or that you've removed all data under the `/data` directory if it's an existing node.

Replacing a Running Node

Occasionally you may need to replace a good node for maintenance purposes. You can replace a node in two ways:

- Add the new node and decommission the node you want to replace.

- Directly replace a node.

I explain both methods in the following sections.

Adding the New Node and Decommissioning the Old Node Afterwards

Replacing a node by adding a new node first is straightforward. Here are the high-level steps for doing this.

1. You add a node as explained in the "Adding a New Node" section.

2. Once you add the new node, decommission the old node as explained in the "Decommissioning a Node" section by running the nodetool decommission command.

3. Run the nodetool cleanup command on all nodes that are part of the same datacenter.

Directly Replacing a Node

The method I explained in the previous section involves streaming data twice or running the nodetool cleanup command. You can simply start the new node with the -Dcassandra.replace_address option to accomplish the same task.

To use this technique, you must ensure that you're not using the consistency level ONE for any keyspace. This means that this node may have the only copy of a row, and thus you risk losing that data.

Here are the steps to directly replace a node.

• First, stop Cassandra on the node you're replacing.

• Follow the steps in the "Replacing a Dead Node" section to replace the node with another node.

If the node you're replacing happens to be a seed node, you must remove the dead node's IP address from the -seeds list in the cassandra.yaml file. You must do this, of course, for all nodes in the cluster. If you need the new node (or a different existing node) to replace the dead seed node, add that node's IP address to the –seeds list.

Removing a Node from a Cluster

If the node you want to remove is healthy and running, just run the nodetool decommission command to remove the node from your cluster. However, if the node is down, you can't decommission it, of course, and that's when you need to run the nodetool removenode command to get rid of the node.

You remove a node with the `nodetool removenode` command by specifying the host ID of the node you wish to remove.

```
$ nodetool removenode 192.168.159.130
```

You can confirm the node removal by running the `nodetool removenode status` command.

```
$ nodetool removenode status
```

Replacing a Dead Node

Nodes can die for many a reason, such as hardware failure. You can tell when a node dies by checking the status of the nodes by running the `nodetool status` command. If you see a status of DN and you don't see any configuration-related errors in the logs, hardware failure or some other issue has crashed the node or is keeping it from starting. In these situations, you must replace the dead node.

Follow these steps to replace a dead node.

1. Add a replacement node to your network.

2. If the dead node happens to be a seed node, replace the IP address of the dead node with that of the new node in the `-seeds` list under the `seed_provider` property.

3. Restart the new node.

4. Install Cassandra on the new node.

5. Configure the `cassandra.yaml` file with the `cluster_name` and the `-seeds` list properties.

6. On the new node, add the rack and datacenter configuration. Assuming you're using the GossipPropertyFileSnitch or the Ec2MultiRegionSnitch, add the rack and datacenter information to the `cassandra-rackdc.properties` file. In addition, remove the `cassandra-topology.properties` file.

7. Start the replacement node with the `replace_address` option, as explained in the "Adding a New Node to a Datacenter" section by either adding the `replace_address` option to the `cassandra-env.sh` file or by starting Cassandra with the `replace_address` option.

Moving a Node to a Different Rack

If you find that you've placed a node in the wrong rack, you can fix the error by moving the node to the correct rack and datacenter. In this case, Cassandra needs to move the data out of the node first, and, after changing the rack the node belongs to, load new data into the node.

The method that DataStax recommends for moving a node is to decommission the node you wish to move to a different rack. Once you decommission it, as explained in the "Decommissioning a Node" section, you can add the node to the rack and datacenter you wish.

Decommissioning an Entire Datacenter

When you decommission a datacenter, you want to first ensure that you don't lose any data by keeping clients from writing to the nodes in this datacenter. Next, implement the following steps to decommission the datacenter.

1. To ensure that all data is propagated to the other datacenters, run the nodetool repair command.

```
$ nodetool repair -full
[2017-07-15 11:50:07,969] Replication factor is 1. No
repair is needed for keyspace 'cycling'
...
[2017-07-15 11:50:08,104] Starting repair command #1
(6c43e1b0-698e-11e7-849e-0920617bbdc2), repairing keyspace
system_traces with repair options (parallelism: parallel,
primary range: false, incremental: false, job threads: 1,
ColumnFamilies: [], dataCenters: [], hosts: [], # of
ranges: 512, pull repair: false)
 [2017-07-15 11:50:12,699] Repair completed successfully
[2017-07-15 11:50:12,710] Repair command #1 finished in 4
seconds
$
```

2. The next step is to delete all references to the datacenter you want to remove. You do this by modifying all the keyspaces in the cluster with the ALTER KEYSPACE command so they don't point to the removed datacenter. The ALTER KEYSPACE command enables you to modify a keyspace's replication strategy, as well as the number of replicas that the database must create in each datacenter.

    As explained in Chapter 4, the SimpleStrategy assigns the same replication factor to the entire cluster and therefore is suitable only for development and testing environments. You must use the NetworkTopologyStrategy option for a production environment. This strategy enables you to specify the replication factor per datacenter.

    For example, if you removed datacenter1, you must ensure that you're configuring a different datacenter for all keyspaces in the cluster, as shown here:

```
ALTER KEYSPACE cycling
WITH REPLICATION = {
    'class'  :  'NetworkTopologyStrategy',
    'datacenter2'  :  3 }
AND DURABLE_WRITES = true ;
```

3. Finally, run the nodetool decommission command on each of the nodes of the datacenter you wish to remove from the cluster.

Switching Snitches

A key point concerning switching of snitches is whether the topology of your cluster will change following a change in the snitch. The replication strategy places replicas on the basis of the information provided by the snitches, since the function of a snitch is to tell Cassandra how to distribute the replicas. When the new snitch you are configuring places replicas in different places, there are topology changes (that is, a change in either the datacenter or rack, or both, where Cassandra places the nodes).

Sometimes there's a change in the network topology and sometimes, not. I address both cases in the following sections.

No Topology Change

If there's no data yet in the cluster, there's no need to change the network topology. You just set the new snitch, and that's that.

For example, you could go from five nodes using a single datacenter using the SimpleSnitch to the same number of nodes in the same data single datacenter using a network snitch (for example, GossipingPropertyFileSnitch).

Here are the steps for switching snitches when there's no topology change.

1. Create the appropriate properties file for denoting the datacenter and rack information depending on the snitch you select, and place the file in the Cassandra configuration directory on all nodes of the cluster. Note that the new snitch isn't enabled yet since the database is running with the snitch previously configured in the `cassandra.yaml` file. The properties file could be one of the following:

   - `cassandra-rackdc.properties` file for the GossipingPropertyFileSnitch, Ec2Snitch,and Ec2MultiRegionSnitch snitch types.

   - `cassandra-topology.properties` file for all the other types of network snitches.

2. Update the value of the `endpoint_snitch` property in the `cassandra.yaml` file on each node of the cluster. In this example, it's

   ```
   endpoint_snitch: GossipingPropertyFileSnitch
   ```

3. Start the nodes one a time, so the changes you've made in the `cassandra.yaml` file regarding the new network snitch type can come into effect.

With Topology Changes

If there's data in the cluster already, a topology change occurs. When the topology changes, new datacenters may be added to the cluster.

Let's consider the case where there's no splitting of datacenters into multiple datacenters (by creating new datacenters). You originally have five nodes in a single cluster, all using the SimpleSnitch. You now change to a new setup with a topology change by going to five nodes in the same single datacenter, but with two racks, using the RackInferrringSnitch.

In this case, the topology changes (from one rack to two racks) but you're not adding any new datacenters. To switch the snitches, first perform the three steps from the previous section. Following this, on each node of the cluster, run the `nodetool repair` (sequential) and the `nodetool cleanup` commands.

If the topology has changes and a new datacenter needs to be added, follow these steps:

1. Create a new datacenter with the new nodes and racks, using the steps in the "Creating a New Datacenter" section.

2. Replicate data into the new datacenter.

3. Remove the nodes from the old datacenters and racks.

4. Run the `nodetool repair` (sequential) and the `nodetool cleanup` commands on each of the cluster's nodes.

Note   If you merely alter the type of snitch and replication, and move some nodes to a new datacenter, you will be replicating data incorrectly.

Managing Gossip

You can monitor and manage gossip with various `nodetool gossip`-related commands.

Getting Information About Gossip

You can get gossip information from the cluster by running the nodetool gossipinfo command, as shown here:

```
$ nodetool  gossipinfo
ubuntu/192.168.159.129
   generation:1500142702
   heartbeat:7498
   STATUS:16:NORMAL,-1270071350005996462
   LOAD:7455:342204.0
   SCHEMA:12:0c912807-68bb-3cf6-91c3-ee14aba78ca6
   DC:8:datacenter1
   RACK:10:rack1
   RELEASE_VERSION:4:3.10
   INTERNAL_IP:6:192.168.159.129
   RPC_ADDRESS:3:192.168.159.129
   NET_VERSION:1:10
   HOST_ID:2:0dbb9e0e-867e-4179-b6b6-631d38dd68f9
   RPC_READY:28:true
   TOKENS:15:<hidden>
/192.168.159.130
  ...
$
```

Disabling and Enabling Gossip

You can effectively take a running node out of the cluster, without stopping the instance, by disabling the gossip protocol on that node.

Execute the nodetool disablegossip command to disable the gossip protocol on a node, as shown here:

```
$ nodetool disablegossip
```

You can review the system.log file to see that Cassandra has marked the node as down, after honoring your request to stop gossip on this node:

```
WARN  [RMI TCP Connection(2)-127.0.0.1] 2017-07-15 13:31:47,640
StorageService.java:318 - Stopping gossip by operator request
INFO  [RMI TCP Connection(2)-127.0.0.1] 2017-07-15 13:31:47,648 Gossiper.
java:1506 - Announcing shutdown
INFO  [RMI TCP Connection(2)-127.0.0.1] 2017-07-15 13:31:47,658
StorageService.java:2248 - Node ubuntu/192.168.159.130 state jump to shutdown
```

If you run the nodetool status command, it shows UN (UP NORMAL) as the node's mode. However, since you turned off gossip, the node isn't in touch with the rest of the cluster.

You can reenable gossip by running the nodetool enablegossip command:

```
$ nodetool enablegossip
```

The log shows that Cassandra has started the gossip process.

```
WARN  [RMI TCP Connection(6)-127.0.0.1] 2017-07-15 13:42:28,187
StorageService.java:331 - Starting gossip by operator request
INFO  [RMI TCP Connection(6)-127.0.0.1] 2017-07-15 13:42:28,197
StorageService.java:2248 - Node ubuntu/192.168.159.130 state jump to NORMAL
WARN  [GossipTasks:1] 2017-07-15 13:42:29,272 FailureDetector.java:288 -
Not marking nodes down due to local pause of 640151266291 > 5000000000
```

Checking the State of Gossip

You can check whether gossip is running by executing the nodetool statusgossip command, as shown here:

```
$ nodetool statusgossip
Running
$
$ nodetool disablegossip
$
$ nodetool statusgossip
not running
$
```

Flushing and Draining Data: The Differences

Both the nodetool flush and the nodetool drain commands enable you to move data from memtables to disk, into SSTables. However, there are differences between the two tools.

Draining a Node

In situations such as when you're getting ready to upgrade Cassandra, you want to ensure that you flush all memtables to disk, into the SSTables. You run the nodetool drain command to flush memtables to SSTables on disk.

You execute the nodetool drain command as shown here:

```
$ nodetool drain
```

The log file shows the following:

```
INFO  [RMI TCP Connection(30)-127.0.0.1] 2017-07-15 13:54:58,272
StorageService.java:1435 - DRAINING: starting drain process
INFO  [RMI TCP Connection(30)-127.0.0.1] 2017-07-15 13:54:58,274
HintsService.java:221 - Paused hints dispatch
INFO  [RMI TCP Connection(30)-127.0.0.1] 2017-07-15 13:54:58,279 Server.
java:176 - Stop listening for CQL clients
INFO  [RMI TCP Connection(30)-127.0.0.1] 2017-07-15 13:54:58,280 Gossiper.
java:1506 - Announcing shutdown
INFO  [RMI TCP Connection(30)-127.0.0.1] 2017-07-15 13:54:58,311
StorageService.java:2248 - Node ubuntu/192.168.159.130 state jump to shutdown
INFO  [RMI TCP Connection(30)-127.0.0.1] 2017-07-15 13:55:00,714
StorageService.java:1435 - DRAINED
```

The nodetool drain command is interesting. When you run this command, Cassandra does other things beyond just flushing the memtables to disk. It also stops listening for any client requests for connections or connection requests from other nodes. You can see this from Cassandra's output when you first drain the node and then try to enable gossip on it.

```
$ nodetool enablegossip
nodetool: Unable to start gossip because the node was drained.
See 'nodetool help' or 'nodetool help <command>'.
$
```

Flushing Data from Memtables

You can also flush memtables to SSTables on disk by running the `nodetool flush` command. You can flush an entire node, or a keyspace, or specific tables. It's a good strategy to flush the memtables to disk prior to taking a snapshot of the database, or one or more keyspaces.

Here's the syntax of the `nodetool flush` command:

```
$ nodetool <options> flush -- <keypsace>  ( <table> ... )
```

You can flush all memtables to disk with the following command:

```
$ nodetool flush
```

You can also flush just one or more keyspaces or tables. Here's how you flush the memtables that belong to the cycling keyspace:

```
$ nodetool flush cycling
```

Unlike the `nodetool drain` command, the `nodetool flush` command just flushes memtables to disk and nothing else. Thus, you're better off using this command instead of the `nodetool drain` command if all you need to do is to flush data to the SSTables.

So, should you flush or drain? The `drain` command, as I explained, stops Cassandra from listening to client requests and requests from the other nodes. You drain the memtables when you want to shut down a node during maintenance and you want the node to start up quicker. Since you flushed the memtables before the restart of the node, the node doesn't need to replay the commitlog following a restart.

Maintaining Datacenters

Common datacenter-related maintenance tasks include adding a datacenter to a cluster, migrating and renaming a cluster without a service interruption, and decommissioning a datacenter.

Adding a Datacenter to a Cluster

Sometimes you many need to add a datacenter to an existing cluster. I show you the steps to do this in this section. When you're done with the process, the old and new datacenters will be replicating to each other.

At the very end, you run the `nodetool rebuild` command, which operates on multiple nodes and streams data from a single source replica to rebuild the token range. This command helps you add the new datacenter to an existing cluster.

Here are the steps for adding a datacenter to a cluster:

1. Ensure that all the current datacenters use NetworkTopologyStrategy as their replication strategy. If they aren't, run the `ALTER KEYSPACE` command to fix things.

Tip   When adding a datacenter to a cluster, don't forget to update the keyspace replication strategies for all keyspaces so they include the new datacenter.

---

2. Install Cassandra software on all nodes of the new datacenter, but don't start the Cassandra service yet.

3. You must next configure some properties in the `cassandra.yaml` file on all the nodes that belong to the new datacenter you're adding. Make sure that important configuration properties such as `-seeds` and `endpoint_snitch` are the same on the new and current nodes.

   In addition, you must also configure the allocation of the vnode tokens to the new nodes. The configuration settings depend on the vnode selection algorithm.

   - *Random Selection Algorithm*: Set the `num_tokens` property (recommended value is 256).

   - *Allocation Algorithm*: Set the `num_tokens` property (recommended value is 8), as well as the `allocate_tokens_for_local_replication_factor` property on all the nodes. The recommended value for the latter property is one of the following:

     - Highest replication factor for a keyspace in this datacenter

     - Replication factor of the keyspace with the heaviest operations

4. Make the following changes on each of the nodes that belong to the new datacenter.

5. Specify the type of snitch in the appropriate properties file. You can't use the SimpleSnitch because that's meant only for a single datacenter, due to its inability to recognize datacenter (or rack) information. Depending on which snitch you choose, you must make changes in the appropriate properties file:

    -   PropertyFileSnitch: `cassandra-topology.properties` file

    -   GossipingPropertyFileSnitch: `cassandra-rackdc.properties` file

6. In the old datacenters, make the following configuration changes:

    -   On some of the existing nodes, add a seed node from the new datacenter. You must restart the nodes on which you make this change since you've modified the `cassandra.yaml` file.

    -   In the appropriate properties file (depends on the type of snitch), specify the new datacenter definition.

7. Start Cassandra on each of the racks, and keep going until you bring up all the nodes.

8. Once all nodes are up, alter the keyspaces as follows:

    ```
    ALTER KEYSPACE "my_ks" WITH REPLICATION =
        {'class': 'NetworkTopologyStrategy', 'ExistingDC':3,
        'NewDC':3};
    ```

9. Finally, run the `nodetool rebuild` command on each node in the new datacenter you just added.

    You can specify several optional parameters when rebuilding nodes, such as `keyspace_name, token_spec,` and `source_dc_name`.

    The `nodetool rebuild` command rebuilds one or a set of keyspaces at a time by streaming data from other nodes. You must specify the name of the keyspace to rebuild via the `keyspace_name` property, as well as `token_spec`, a property that enables you to specify a single token, a list of single tokens, or a range of tokens (`start_token, end_token`).

You can specify the `source-dc-name` with the `nodetool rebuild` command. This attribute refers to the name of the datacenter that Cassandra uses as the source for streaming. Cassandra can build from any datacenter, and if you omit the name of the source datacenter, it randomly chooses a datacenter.

If the rebuild fails for any reason, you can restart it, at which time the process resumes from where it had stopped. You can also do a selective rebuild by specifying the `-ts` or `-token` options to specify either a list of tokens or a token range (or ranges).

The simplest invocation of the `nodetool rebuild` command is

```
$ nodetool rebuild
```

Decommissioning a Datacenter

Decommissioning a datacenter is the same as removing a datacenter. You can decommission a datacenter without losing any data by following these steps.

1. Before you start decommissioning the datacenter, ensure that clients aren't writing to any nodes in the datacenter. You can confirm that there are no pending write requests in any node by running the following command:

    ```
    $ nodetool tpstats
    ```

2. Propagate the data from the datacenter you're planning to decommission by running a full repair with the following command:

    ```
    $ nodetool repair -full
    ```

3. Alter all keyspaces in the database to make sure that none of them references the datacenter you're decommissioning.

    Let's say you've three datacenters: DC1, DC2, and DC3. You're removing datacenter DC3 from your cluster. You must therefore remove datacenter DC3 from all keyspace configurations. If you've set a replication factor for DC3 for any keyspace, run the ALTER KEYSPACE command, as shown here, to remove the datacenter DC3 from that keyspace's configuration:

```
cqlsh> alter keyspace cycling WITH replication = {'class':
       'NetworkTopologyStrategy', 'DC1':1,'DC2':2};
```

4. Run the nodetool decommission command on all nodes of the datacenter you're decommissioning.

   ```
   $ nodetool decommission
   ```

5. Once the decommissioning is completed, shut the node down. You can confirm that all nodes in the datacenter were removed by running the nodetool status command.

Handling Data Corruption

Cassandra offers tools to check for corrupted data and to fix the data corruption by rebuilding the table with the corrupted data.

The following sections show you how to detect data corruption and to rebuild corrupted tables.

Checking for Data Corruption

Run the sstableverify command to check a specific SSTable for errors or corrupted data. The following is an example where *cycling* is the name of the keyspace and cycle is the name of the table you want to check out:

```
$ sstableverify --verbose cycling cycle
Verifying BigTableReader(path='/cassandra/apache-cassandra-3.10/data/
data/cycling/cycle-2276fb7064e911e7b186d794a4e00229/mc-2-big-Data.db')
(0.029KiB)
...
Checking computed hash of BigTableReader(path='/cassandra/apache-
cassandra-3.10/data/data/cycling/cycle-2276fb7064e911e7b186d794a4e00229/
mc-3-big-Data.db')
$
```

Fixing Corrupted Data

You can fix corrupted data by rebuilding the SSTables. Rebuilding the SSTable will remove the corrupted data, leaving the useful data intact. There are two tools that help you rebuild tables: the nodetool scrub command and the *sstablescrub* utility. These two tools are quite similar. Your first option is the nodetool scrub command.

Removing Corrupt Data by Rebuilding the Table

You can remove corrupted data with the help of the nodetool scrub command. This command rebuilds the SSTables for one or more tables.

The following is the syntax for the nodetool scrub command:

```
$ nodetool <options> scrub <keyspace>
--  -ns | --no-snapshot
-s | --skip-corrupted    <table> ...
```

Here's an example where you scrub the tables in the cycling tablespace:

```
INFO  [CompactionExecutor:50] 2017-07-16 11:47:14,560 OutputHandler.
java:42 - Scrubbing BigTableReader(path='/cassandra/apache-cassandra-3.10/
data/data/cycling/cyclist_name-4d1743b060ef11e7805be14006afbdda/mc-1-big-
Data.db') (0.206KiB)
INFO  [CompactionExecutor:50] 2017-07-16 11:47:14,604 OutputHandler.
java:42 - Scrub of BigTableReader(path='/cassandra/apache-cassandra-3.10/
data/data/cycling/cyclist_name-4d1743b060ef11e7805be14006afbdda/mc-1-big-
Data.db') complete; looks like all 0 rows were tombstoned
...
```

The nodetool scrub command rebuilds SSTables and when it does so, it discards data that may be corrupt. To safeguard the data, the command also snapshots the data files of the SSTable before rebuilding the table.

The following are the key options for the nodetool scrub command:

- --no-snapshot: This option will disable the default snapshot creation for the SSTables that Cassandra will rebuild. Specifying this option saves storage space and means you don't have to remove the snapshots later.

- `--skip-corrupted`: This option will let the rebuild operation skip corrupted partitions where the operation can't validate a column's values against that column's data type.

Offline Rebuilding of a Table

The *sstablescrub* utility helps you rebuild a table when the node is offline. It works the same as the `nodetool scrub` command. If the `nodetool scrub` command fails to rebuild the table with the corrupted data, then your next option should be to run the *sstablescrub* utility.

Since *sstablescrub* is an offline utility, you must first shut down the node. Following this, run the following command to scrub the table:

```
$ sstablescrub cycling cycle
```

Managing Handoff and Hints

You can use the nodetool utility to manage several aspects of the handoff process and the hints mechanism. By running the `nodetool` commands, you can enable and disable handoff, pause and resume handoff, and disable and enable hints for a specific datacenter. Here's a listing of the handoff- and hints-related `nodetool` commands:

- `nodetool disablehandoff`: Disables the storing of hints on a node
- `nodetool enablehandoff`: Enables the storing of hints on a node
- `nodetool pausehandoff`: Pauses the dispatch of hints
- `nodetool resumehandoff`: Resumes the dispatch of hints

The `nodetool enablehintsfordc` command enables you to turn on hints for a datacenter. You do this when you've blacklisted a datacenter with the `hnted_handoff_disabled_datacenters` property in the `cassandra.yaml` file, or when you've disabled hints for a datacenter with the `disablehintsfordc` command for a datacenter.

Here's how you toggle hints on and off for a datacenter:

```
$ nodetool enablehintsfordc DC2
$ nodetool disablehintsfordc DC2
```

You disable hints for a datacenter if a datacenter is down, or when you're failing a datacenter over. The database will continue to send the hints to the other datacenters.

Purging the Gossip State on a Node

Cassandra stores the gossip information on each node so the nodes can use it when restarting. If the nodes need to retrieve the gossip data from elsewhere in the cluster, node startups will be slower. Occasionally, you may need to manually fix problems with gossip when a node doesn't have the correct state of gossip.

Follow these steps to fix problems regarding an incorrect gossip state.

1. Shut down the node experiencing the gossip problems.

   ```
   $ nodetool assassinate
   ```

2. Stop the client applications from writing to the Cassandra cluster.

3. Take all nodes offline, first draining each of the nodes.

   ```
   $ nodetool drain
   $ sudo service cassandra stop
   ```

4. Remove all directories in the peers-UUID directory.

   ```
   $ sudo rm -r /var/lib/cassandra/data/system/peers-UUID/8
   ```

5. After removing the data from the peers directory, run the following SQL statement on each of the nodes to ensure that the nodes can see each other:

   ```
   cqlsh> select * from system.peers;
   ```

6. Set the cassandra.load_ring_state=false property on each node to clear the gossip state when you restart the nodes in the next step. You can set this property from the command line, or by adding the property in the cassandra-env.sh file, as explained (for a different parameter) in the "Decommissioning Nodes" section in this chapter.

7. Restart all nodes in the cluster. Make sure you undo any changes to the cassandra-env.sh file that you might have made in Step 6.

Summary

I explained several important cluster administration tasks in this chapter. The nodetool utility helps enormously when removing nodes, replacing dead nodes, moving nodes, assassinating nodes, and removing nodes from a ring. You also learned how to decommission and add datacenters.

The best way to gain confidence in performing the cluster management tasks is to practice those tasks in a test cluster. You're likely to run into minor issues when performing some of the changes, and practicing tasks such as decommissioning nodes and datacenters teaches you how to fix those issues quickly, thus saving time during changes in a production setting.

CHAPTER 10

Monitoring, Logging, and Metrics

Monitoring Cassandra is in many ways similar to how you manage a traditional relational database. The complexity lies in the fact that Cassandra is a distributed database, so you'll be concerning yourself also with the distribution of data and the balancing of the workload among the nodes of a cluster.

The nodetool utility offers many commands that help you monitor your cluster, and you learned many of these commands in the earlier chapters.

JConsole is a powerful monitoring tool that you can use both from the command line and as a GUI tool.

Cassandra offers numerous metrics that help you evaluate and monitor a cluster's performance and health. These metrics include table and keyspace metrics, as well as cache and client request metrics, for example. This chapter reviews important Cassandra metrics.

In the final section of this chapter, I show how to set up and configure Nagios, which is a popular tool that helps monitor Cassandra clusters.

The nodetool proxyhistograms Command

The nodetool proxyhistograms command shows the network statistics in a cluster.

```
$ nodetool proxyhistograms
proxy histograms
Percentile          Read Latency      Write Latency       Range Latency
CAS Read Latency  CAS Write Latency View Write Latency
                    (micros)            (micros)              (micros)
(micros)           (micros)            (micros)
```

© Sam R. Alapati 2018
S. R. Alapati, *Expert Apache Cassandra Administration*, https://doi.org/10.1007/978-1-4842-3126-5_10

50%		943.13		0.00		4866.32
	0.00		0.00		0.00	
75%		4055.27		0.00		14530.76
	0.00		0.00		0.00	
95%		4055.27		0.00		20924.30
	0.00		0.00		0.00	
98%		4055.27		0.00		20924.30
	0.00		0.00		0.00	
99%		4055.27		0.00		20924.30
	0.00		0.00		0.00	
Min		785.94		0.00		1358.10
	0.00		0.00		0.00	
Max		4055.27		0.00		20924.30
	0.00		0.00		0.00	

```
$
```

Getting Table-Level Statistics

Run the nodetool tablestats (formerly nodetool cfstats) command to get statistics about one or more tables. Be sure to specify the keyspace and the table name, as I show in the following example. By default, Cassandra will spit out statistics about all tables!

Cassandra updates the table statistics when you flush data or when the database changes the SSTables by compacting them.

You can have Cassandra ignore some tables by specifying the table names with the -I flag. The following example shows how to get stats for a table:

```
$ nodetool tablestats test.kv2
Total number of tables: 40
----------------
Keyspace : test
        Read Count: 0
        Read Latency: NaN ms.
        Write Count: 0
        Write Latency: NaN ms.
        Pending Flushes: 0
                Table: kv2
```

```
SSTable count: 1
Space used (live): 5012
Space used (total): 5012
Space used by snapshots (total): 0
Off heap memory used (total): 194
SSTable Compression Ratio: 1.2
Number of keys (estimate): 1
Memtable cell count: 0
Memtable data size: 0
Memtable off heap memory used: 0
Memtable switch count: 0
Local read count: 0
Local read latency: NaN ms
Local write count: 0
Local write latency: NaN ms
Pending flushes: 0
Percent repaired: 0.0
Bloom filter false positives: 0
Bloom filter false ratio: 0.00000
Bloom filter space used: 176
Bloom filter off heap memory used: 168
Index summary off heap memory used: 18
Compression metadata off heap memory used: 8
Compacted partition minimum bytes: 30
Compacted partition maximum bytes: 35
Compacted partition mean bytes: 35
Average live cells per slice (last five minutes): NaN
Maximum live cells per slice (last five minutes): 0
Average tombstones per slice (last five minutes): NaN
Maximum tombstones per slice (last five minutes): 0
Dropped Mutations: 0
$
```

As you can see, the nodetool tablestats command produces exhaustive statistics about a table. Besides space usage statistics, the command also reveals read and write latency statistics and Bloom filter-related statistics.

Getting Network Information from the Host

You can get network information about a node by running the nodetool netstats command. The output of this command shows information such as the operational mode of the node (NORMAL, DECOMMISSIONED, LEAVING, etc.), read repair statistics, and the number of inactive, pending, and completed commands plus their responses.

Here's an example:

```
$ nodetool netstats -H
Mode: NORMAL
Not sending any streams.
Read Repair Statistics:
Attempted: 0
Mismatch (Blocking): 0
Mismatch (Background): 0
Pool Name                      Active    Pending       Completed   Dropped
Large messages                   n/a          1               0         0
Small messages                   n/a          0            1058         0
Gossip messages                  n/a          0           18589         0
$
```

By default, Cassandra will show the results for the node from which you issue this command, but it allows you to specify a remote node with the -h option.

The nodetool tablehistograms Command

The nodetool tablehistograms (formerly nodetool cfhistograms) command provides statistics about tables that you can use to plot a frequency function.

This command is not cumulative; it monitors only the operations since you last ran the command in the current session.

The following is an example showing how to run the nodetool tablehistograms command. In this command, *cycling* refers to the keyspace and cycle to the table.

```
$ nodetool tablehistograms cycling cycle
cycling/cycle histograms
```

Percentile Size	SSTables Cell Count	Write Latency (micros)	Read Latency (micros)	Partition	(bytes)
50%	0.00	0.00	0.00	35	1
75%	0.00	0.00	0.00	35	1
95%	0.00	0.00	0.00	35	1
98%	0.00	0.00	0.00	35	1
99%	0.00	0.00	0.00	35	1
Min	0.00	0.00	0.00	30	0
Max	0.00	0.00	0.00	35	1

$

Here's a description of the key columns in the output of the command:

- Percentile: The percentile rank

- SSTables: The number of SSTables accessed per read during a recent read

- Write Latency: The write latency in microseconds for recent writes

- Read Latency: The read latency in microseconds for recent reads

- Partition Size: The partition size in bytes

Checking the Cluster Health

The nodetool utility is your best friend when it comes to monitoring a cluster and performing routine maintenance tasks.

Key nodetool commands that you'll often use for checking a cluster's health are the following:

- nodetool status

- nodetool info

- nodetool tpstats

In the following sections, I review these key nodetool cluster health monitoring tools.

The nodetool status Command

You've seen the nodetool status command in action numerous times in this book. This command enables you to check the health of a cluster's nodes. In addition, it also lets you know about the distribution of data among the nodes. Monitor the cluster with this command, and if it shows an unbalanced cluster due to too many nodes in a rack, move some of the nodes around using the techniques explained in the previous chapter.

The nodetool info Command

Run the nodetool info command to get node information, such as disk load, uptime, and heap memory usage. The command also provides valuable information about how the database is utilizing all three of its caches: the key cache, the row cache, and the counter cache.

The nodetool info command provides valuable information pertaining to the nodes, such as the following:

- Disk storage (load) information

- Times started (generation)

- Uptime

- Heap memory usage

- Key, row, counter, and chunk cache information

- Status of gossip (active or not)

- Percent repaired

- Token information (optionally)

Here's an example:

```
$ nodetool info
ID                     : 7d9e3ac3-ffdc-4cee-be82-4b0c6e614aa1
Gossip active          : true
Thrift active          : false
Native Transport active: true
Load                   : 394.92 KiB
Generation No          : 1500219460
```

```
Uptime (seconds)         : 8441
Heap Memory (MB)         : 95.36 / 1014.00
Off Heap Memory (MB)     : 0.00
Data Center              : datacenter1
Rack                     : rack1
Exceptions               : 0
Key Cache                : entries 32, size 2.7 KiB, capacity 50 MiB, 159
                           hits, 200 requests, 0.795 recent hit rate, 14400
                           save period in seconds
Row Cache                : entries 0, size 0 bytes, capacity 0 bytes, 0 hits,
                           0 requests, NaN recent hit rate, 0 save period in
                           seconds
Counter Cache            : entries 0, size 0 bytes, capacity 25 MiB, 0 hits,
                           0 requests, NaN recent hit rate, 7200 save period
                           in seconds
Chunk Cache              : entries 17, size 1.06 MiB, capacity 221 MiB, 58
                           misses, 291 requests, 0.801 recent hit rate,
                           497.940 microseconds miss latency
Percent Repaired         : 100.0%
Token                    : (invoke with -T/--tokens to see all 256 tokens)
$
```

Using Thread Pools Statistics (nodetool tpstats)

The nodetool tpstats command shows the usage statistics of thread pools. Cassandra breaks tasks into stages, with each stage using a separate queue and a thread pool. A messaging service connects the various stages.

The nodetool tpstats command provides information about each stage of an operation by thread pool. It shows the following things:

- Number of active threads

- Number of requests waiting to be executed by a thread pool

- Number of tasks a thread pool has completed

- Number of blocked requests due to the thread pool in the next step being full

- Number of total blocked requests in this thread pool up until this point in time

The database refreshes the information that the `nodetool tpstats` command provides when you flush memtables or when the database compacts any SSTables.

Here's an example showing how to run the `nodetool tpstats` command:

```
$ nodetool tpstats
```

Pool Name	Active	Pending	Completed	Blocked	All time blocked
ReadStage	0	0	0	0	0
MiscStage	0	0	0	0	0
CompactionExecutor	0	0	5532	0	0
MutationStage	0	0	1028	0	0
MemtableReclaimMemory	0	0	29	0	0
endingRangeCalculator	0	0	2	0	0
GossipStage	0	0	26038	0	0
SecondaryIndexManagement	0	0	0	0	0
HintsDispatcher	0	0	0	0	0
RequestResponseStage	0	0	1031	0	0
Native-Transport-Requests	0	0	0	0	0
ReadRepairStage	0	0	0	0	0
CounterMutationStage	0	0	0	0	0
MigrationStage	0	0	0	0	0
MemtablePostFlush	0	0	73	0	0
PerDiskMemtableFlushWriter_	0	0	0	29	0
ValidationExecutor	0	0	2	0	0
Sampler	0	0	0	0	0
MemtableFlushWriter	0	0	29	0	0
InternalResponseStage	0	0	2	0	0
ViewMutationStage	0	0	0	0	0
AntiEntropyStage	0	0	8	0	0
CacheCleanupExecutor	0	0	0	0	0

```
Message type Dropped
READ                    0
RANGE_SLICE             0
```

```
_TRACE                    0
HINT                      0
MUTATION                  0
COUNTER_MUTATION          0
BATCH_STORE               0
BATCH_REMOVE              0
REQUEST_RESPONSE          0
PAGED_RANGE               0
READ_REPAIR               0
$
```

The output of the nodetool tpstats command shows statistics pertaining to specific thread pools associated with tasks in the database. All this detailed information is quite useful to you when you're troubleshooting issues or tuning the database. You can determine strategies to fix performance issues based on the activity in the pools. High numbers in a pool point to symptoms of problems in the database.

The following is a brief list of the thread pools, the tasks associated with those pools, and the actions you can perform to improve things.

- *AntiEntropyStage*: This pool processes repair messages. You can use the nodetool repair command to perform repairs.

- *GossipStage*: Distributes the node information. You may see issues in this pool due to some schemas being out of sync with each other. Use the nodetool resetlocalschema command to sync the schemas.

- *HintedHandoff*: Sends missed changes such as updates and deletes to other nodes. Sometimes you'll see issues with the handoff process due to various reasons. You can use the nodetool disablehandoff and the nodetool repair commands to fix handoff related issues.

- *MutationStage*: This stage performs inserts and updates on the local node and replays the commitlog. Hints in progress are also part of this stage. If you see a high number of pending write requests, it may indicate that the node is overloaded. You can add nodes or rewrite code to reduce the mutations.

- ReadRepairStage: This stage shows the waits for performing read repairs. If the Pending numbers are high, you can lower the read_repair _chance value for tables that are read often, to value such as 0.11. The read_ repair_chance attribute for a table is the probability that a successful read operation will trigger a read repair (the default value is 0.0).

You can monitor the thread pool statistics closely, and if you see a consistent increase in the Pending tasks column values, you know it's time to add additional capacity to the cluster.

At the bottom of the output for the nodetool tpstats command is a section for dropped messages for this node, meaning that the node received more messages than it could handle. If you see a significant number of blocked tasks and/or dropped messages, it means the database is having trouble keeping up with the current workloads.

Using JMX Clients

Cassandra exposes numerous metrics and management operation commands over Java Management Extensions (JMX). JMX provides tools for monitoring and managing Java applications and services. It can monitor any statistics and manage any operations that Java applications expose as MBeans (management beans).

JMX also enables you to remotely connect to a cluster's instances so you can monitor Cassandra performance (memory, threads, CPU usage, etc.) and manage Cassandra by helping you modify certain runtime properties. MBeans are special JavaBeans that enable you to access resources inside the JVM externally. Through JMX, you can programmatically review settings for various entries such as memory, threads, CPU, Gossip, and other Cassandra-related components that are instrumented in JMX.

Under the covers, nodetool commands that you execute access JMX metrics to do their work. Nodetool supports key JMX metrics and operations, as well as additional Cassandra administration-related commands (such as the proxyhistograms command). However, nodetool can't access some metrics, and in those cases, you can use a general purpose JMX client such as JConsole or jmxsh, as I explain in the following sections.

JConsole

JConsole is a JMX client that captures JMX metrics and operations exposed by Cassandra and displays them in a graphical format. JConsole is a standard tool for managing MBeans. JConsole is at times hard to use with remote servers because you must have ports open in the firewall for its connections. This tool is thus not ideal for a production environment, although it's great for development and test servers.

Tip   Since JConsole consumes a lot of resources, DataStax recommends that you run the tool on a node that doesn't host a Cassandra instance.

JConsole provides information about memory and thread usage, Java class loading, and other information about the Java Virtual machine (JVM), and MBeans. You can use the Memory tab to perform a Java garbage collection.

A key metric offered by the JConsole is compaction metrics. By monitoring compaction metrics, you can figure out when to add capacity to a cluster.

You can connect to a Cassandra instance through JConsole. To do this, first start up JConsole.

```
# /usr/lib/jvm/java-8-oracle/bin/jconsole
```

You'll see the *jconsole:New Connection* page appear, as shown in Figure 10-1. Select the Remote process button and enter the address of the Cassandra node and the port (in my case, 102.168.19.129:7199).

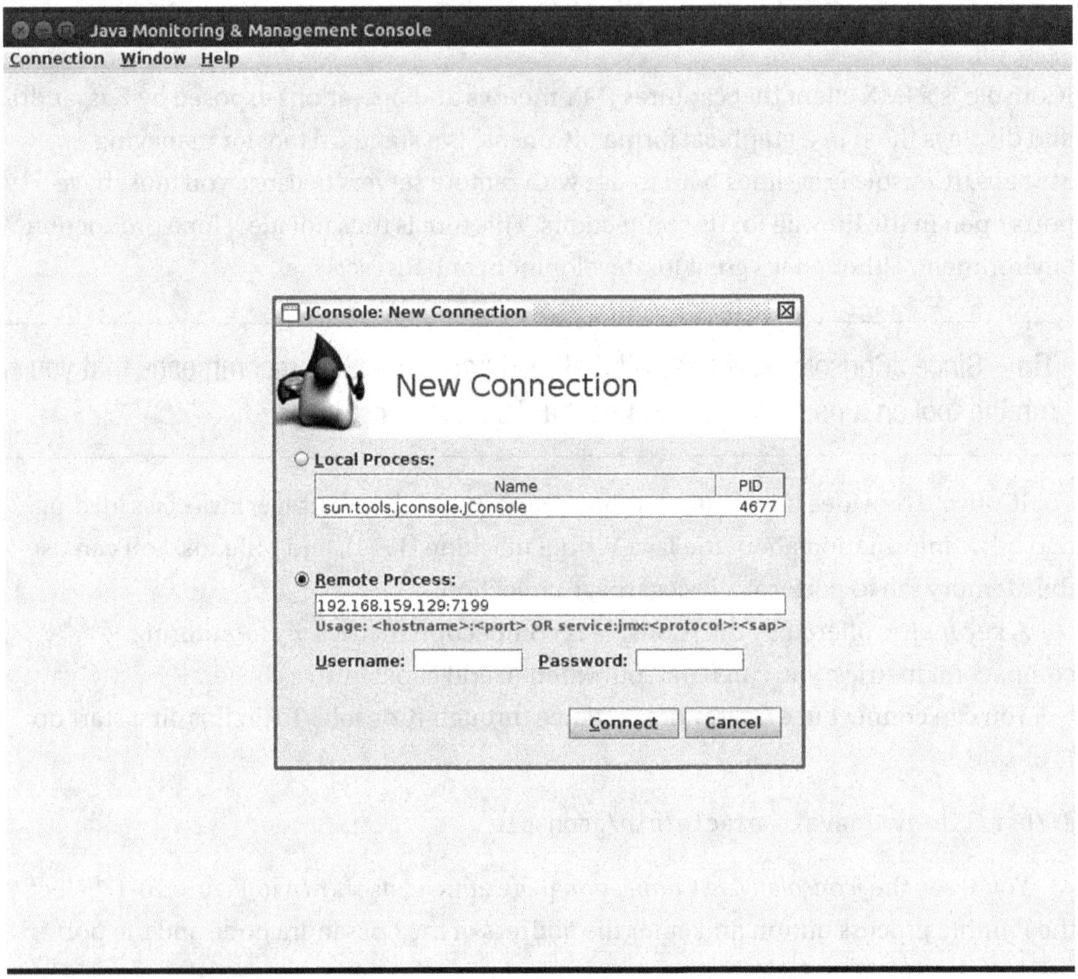

Figure 10-1. *The JConsole New Connection login page*

Once you log in, you'll see the following tabs: Overview, Memory, Threads, Classes, VM Summary, and MBeans. Figure 10-2 shows the Overview tab. Each of the tables presents a set of graphs that offer a snapshot of various resources in the Cassandra node.

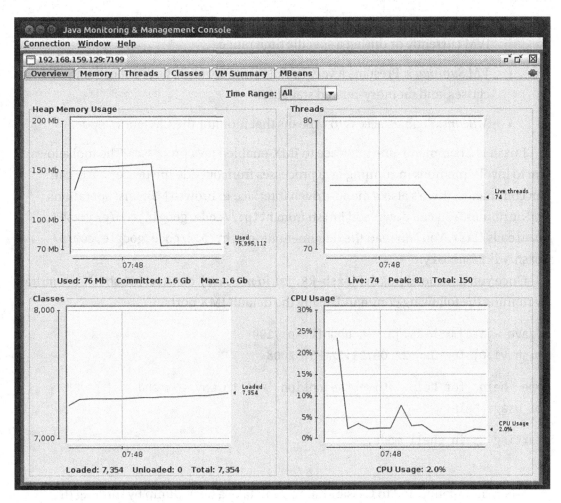

Figure 10-2. *The JConsole Overview tab*

The following is a description of the various JConsole tabs:

- *Overview*: Presents a set of four graphs for tracking CPU, memory, threads, and classes.

- *Memory*: Shows the current state of the Java heap usage. You can also view garbage collection metrics from this tab and manually start garbage collection from here.

- *Threads*: Shows the current and peak usage pattern of various thread stages, such as compaction and garbage collection.

- *Classes*: Presents a graph showing how many classes are loaded in the JVM currently or during a specific time range.

- *VM Summary*: Presents a very useful summary of garbage collection, classes, and memory-related statistics.

- *MBeans*: Enables access to MBeans that monitor the Cassandra node.

Jmxsh is a command-line interface to JMX-enabled Java processes. The tool allows you to invoke methods in running Java processes from outside the process. You can run commands; there's also a menu-driven interface to browse MBeans, operations, and attributes. You can download jmxsh from http://code.google.com/p/jmxsh/downloads/list. You can read the documentation at http://code.google.com/p/jmxsh/wiki/Summary.

Once you download jmxsh (jmxsh-R5.jar in my case), connect jmxsh to Cassandra by running the following command, using the default JMX port:

```
$ java -jar jmxsh-R5.jar -h ubuntu -p 7199
jmxsh v1.0, Tue Jan 22 08:23:12 PST 2008

Type 'help' for help.  Give the option '-?' to any command
for usage help.

Starting up in shell mode.
%
```

You're now connected to Cassandra. You can take a heap dump by running the following jmxsh command:

```
% jmx_invoke -m com.sun.management:type=HotSpotDiagnostic dumpHeap /tmp/
heapdump.hprof false
```

You can bulk load Cassandra tables by running the following command:

```
% jmx_invoke -m org.apache.cassandra.db:type=StorageService bulkLoad /path/
to/SSTables
```

jmxsh also offers a browser mode. You enter the browser mode by pressing enter at the jmxsh prompt:

```
Entering browse mode.
========================================================

Available Domains:

        1. java.util.logging
        2. org.apache.cassandra.service
        3. org.apache.cassandra.db
        4. org.apache.cassandra.metrics
        5. java.nio
        6. org.apache.cassandra.hints
        7. JMImplementation
        8. java.lang
        9. com.sun.management
       10. ch.qos.logback.classic
       11. org.apache.cassandra.internal
       12. org.apache.cassandra.request
       13. org.apache.cassandra.net

  SERVER: service:jmx:rmi:///jndi/rmi://ubuntu:7199/jmxrmi


========================================================
Select a domain:
```

With this introduction to JMX clients under your belt, let's turn to how you can manage Cassandra's logging capabilities.

Logging

There are two main Cassandra logs: system.log and debug.log. The system.log file logs all activity in the database, including the startup and shutdown activity, as well as all node-related tasks activity. It also logs all changes you make to the database schemas such as modifications to keyspaces and tables.

Logging in Cassandra uses the Simple Logging façade for Java (SLF4J) with a logback backend. You can configure logging programmatically or manually. In this chapter, I focus on the manual configuration of logging, which you can do through one of the following ways.

- By configuring the `logback.xml` file. By default, logback looks for the `logback-test.xml` and the `logback.xml` files, in that order, for logging configuration details.

- You can configure the logging levels by running the `nodetool setlogginglevel` command.

- You can also configure logging through a JMX client such as the JConsole tool.

The following sections explain how to configure logging using all three methods. But first, let me explain how to set the logging locations where Cassandra stores the various logging files.

Setting the Logging Locations

The installation method for Cassandra determines the default logging location. You may have the default log location as `/var/log/cassandra` or as the `$CASSANDRA_HOME/logs` directory, depending on whether you used a package or tarball-based installation method. It's in the subdirectory named `logs` that you'll find the two important Cassandra log files: `system.log` and `debug.log`.

You can change the default logging location by adding the following line to the `cassandra-env.sh` file:

```
export CASSANDRA_LOG_DIR=/new/log/location
```

Setting the `CASSANDRA_LOG_DIR` parameter in the `cassandra-env.sh` file will make Cassandra store all of its log files in this location. If you want to specify a different directory for the `system.log` and `debug.log` files, for example, you can do so by specifying the locations in the `logback.xml` file, through configuring the `${cassandra.logdir}` attribute for the two log files.

Note   You can also configure logging through JMX with the JConsole tool.

You can also modify the location of the log directory by editing the following entry in the `logback.xml` file and specifying a different filename:

```
<file>${cassandra.logdir}/system.log</file>
```

Configuring Logging Through Logback

Cassandra's logging capabilities are supported by SLF4J with a logback backend. You can manually configure logging by configuring the `logback.xml` (or the `logback-test.xml`) file that's automatically installed with Cassandra.

Since logback is a critical part of Cassandra logging, I review the essentials of the logback framework in the following section.

The Logback Logging Framework

Logback is a successor to the well-known log4j logging framework. Logback was created by the same person who created log4j and it builds on the previous work on log4j, which has been highly successful as a logging framework. As the creator of Logback says: "If you like log4j, you'll love Logback."

The Logback framework offers many benefits over log4j, such as

- An extensive battery of tests, leading to a highly dependable logging framework

- Automatic reloading of configuration files after you modify them

- Automatic compression of archived log files during rollover

- Automatic removal of old log archives

- Conditional processing of configuration files using if-then constructs, so a single configuration file can serve multiple environments

- The SiftingAppender is highly versatile, and you can use it to separate logging according to various runtime attributes, such as user sessions.

In the following section, I briefly describe how the Logback logging framework works.

Main Components of the Logback Framework: Logger, Appenders, and Layouts

Logback uses three main components that work together to log messages according to message types and level, and control the formatting of the messages and where it reports the log messages. These three main classes are the following:

- `Logger`
- `Appender`
- `Layout`

Logger

The `logger` class determines what Logback will log. It allows Logback to selectively enable or disable logging requests based on their logger. The root logger is at the top of the hierarchy and is part of every hierarchy. Loggers may be assigned levels such as TRACE, DEBUG, ERROR, etc. A logger will inherit its log level from its ancestor if you don't assign it one. The root logger always has an assigned level, which is DEBUG by default. This ensures that all child loggers of the root logger will inherit a level.

Logback orders its logging levels as follows:

`TRACE < DEBUG < INFO <  WARN < ERROR`

If you set the log level of the root logger to `Level.OFF`, the highest log level, you'll be completely turning off logging.

Appender

An appender is responsible for writing a logging event to components, and it points to one or more output destinations where Logback allows logging requests to print. For example, a logging request can be printed to the console (appender name='CONSOLE"), files (`<appender name="FILE" class="ch.qos.logback.core.FileAppender">`), and databases such as MYSQL and Oracle, as well as remote Linux Syslog daemons.

You can configure more than one appender to a logger. Appenders inherit additively from the logger hierarchy, meaning that if you add a console appender to the root logger, then all logging requests will show up on the console.

The special appender `RollingFileAppender` has the capability to roll over log files. This appender can, for example, log to file named `mylog.txt`, and after it satisfies some conditions that you set, change its logging target over to a different file.

You can add *filters* to an appender to filter events by criteria such as the contents of the log message or the time of the day.

Layout

Whereas an appender helps you customize the output destination, the `layout` class customizes the output format. You do this by associating a layout with an appender.

A layout configures the formatting of the logging request, and the appender is responsible for sending the formatted output to the destinations you configure.

The logback.xml File

The following is a sample `logback.xml` file showing various types of appenders:

```
<configuration scan="true">
  <jmxConfigurator />
  <!-- No shutdown hook; we run it ourselves in StorageService after
  shutdown -->
  <!-- SYSTEMLOG rolling file appender to system.log (INFO level) -->
  <appender name="SYSTEMLOG" class="ch.qos.logback.core.rolling.Rolling
  FileAppender">
    <filter class="ch.qos.logback.classic.filter.ThresholdFilter">
      <level>INFO</level>
    </filter>
    <file>${cassandra.logdir}/system.log</file>
    <rollingPolicy class="ch.qos.logback.core.rolling.FixedWindow
    RollingPolicy">
      <fileNamePattern>${cassandra.logdir}/system.log.%i.zip</fileName
      Pattern>
      <minIndex>1</minIndex>
      <maxIndex>20</maxIndex>
    </rollingPolicy>
    <triggeringPolicy class="ch.qos.logback.core.rolling.SizeBased
    TriggeringPolicy">
```

```
      <maxFileSize>20MB</maxFileSize>
    </triggeringPolicy>
    <encoder>
      <pattern>%-5level [%thread] %date{ISO8601} %F:%L - %msg%n</pattern>
    </encoder>
  </appender>
  <!-- DEBUGLOG rolling file appender to debug.log (all levels) -->
  <appender name="DEBUGLOG" class="ch.qos.logback.core.rolling.Rolling
  FileAppender">
    <file>${cassandra.logdir}/debug.log</file>
    <rollingPolicy class="ch.qos.logback.core.rolling.FixedWindow
    RollingPolicy">
      <fileNamePattern>${cassandra.logdir}/debug.log.%i.zip</
      fileNamePattern>
      <minIndex>1</minIndex>
      <maxIndex>20</maxIndex>
    </rollingPolicy>
    <triggeringPolicy class="ch.qos.logback.core.rolling.SizeBased
    TriggeringPolicy">
      <maxFileSize>20MB</maxFileSize>
    </triggeringPolicy>
    <encoder>
      <pattern>%-5level [%thread] %date{ISO8601} %F:%L - %msg%n</pattern>
    </encoder>
  </appender>
  <!-- ASYNCLOG assynchronous appender to debug.log (all levels) -->
  <appender name="ASYNCDEBUGLOG" class="ch.qos.logback.classic.
  AsyncAppender">
    <queueSize>1024</queueSize>
    <discardingThreshold>0</discardingThreshold>
    <includeCallerData>true</includeCallerData>
    <appender-ref ref="DEBUGLOG" />
  </appender>
  <!-- STDOUT console appender to stdout (INFO level) -->
  <appender name="STDOUT" class="ch.qos.logback.core.ConsoleAppender">
```

```xml
    <filter class="ch.qos.logback.classic.filter.ThresholdFilter">
      <level>INFO</level>
    </filter>
    <encoder>
      <pattern>%-5level [%thread] %date{ISO8601} %F:%L - %msg%n</pattern>
    </encoder>
  </appender>

  <!-- Uncomment bellow and corresponding appender-ref to activate logback
  metrics
  <appender name="LogbackMetrics" class="com.codahale.metrics.logback.
  InstrumentedAppender" />
   -->
  <root level="INFO">
    <appender-ref ref="SYSTEMLOG" />
    <appender-ref ref="STDOUT" />
    <appender-ref ref="ASYNCDEBUGLOG" /> <!-- Comment this line to disable
    debug.log -->
    <!--
    <appender-ref ref="LogbackMetrics" />
    -->
  </root>
  <logger name="org.apache.cassandra" level="DEBUG"/>
  <logger name="com.thinkaurelius.thrift" level="ERROR"/>
</configuration>
```

Note   You can use either .xml or .groovy as the file extension for the Logback configuration file. In our case, I use .xml.

When you need to dig deep into the internals of Cassandra, one of the best things to do is to modify the logging level to a more "liberal" level, meaning one that shows more details about the operations Cassandra is performing (or failing to perform!).

The default logging level of INFO doesn't generate many details about operations, simply providing the status of operations, as shown here:

```
WARN  [main] 2017-04-03 13:06:55,973 DatabaseDescriptor.java:493 - Small
cdc volume detected at ./../data/cdc_raw; setting cdc_total_space_in_mb to
2377.  You can override this in cassandra.yaml
INFO  [main] 2017-04-03 13:06:56,882 CassandraDaemon.java:472 - Hostname:
ubuntu
INFO  [main] 2017-04-03 13:06:56,887 CassandraDaemon.java:479 - JVM vendor/
version: Java HotSpot(TM) 64-Bit Server VM/1.8.0_121
INFO  [main] 2017-04-03 13:06:56,897 CassandraDaemon.java:480 - Heap size:
476.000MiB/476.000MiB
INFO  [main] 2017-04-03 13:06:56,911 CassandraDaemon.java:485 - Par
Eden Space Heap memory: init = 83886080(81920K) used = 77260784(75449K)
committed = 83886080(81920K) max = 83886080(81920K)
```

You can change the default logging level to DEBUG by modifying the logback.xml file as follows:

```
<root level="DEBUG">
        <appender-ref ref="FILE" />
        <appender-ref ref="STDOUT" />
</root>
```

You don't have to restart Cassandra after modifying the logback.xml file because the database scans the logback.xml file every minute looking for changes, provided you've added the following line to the file:

```
<configuration scan="true">
```

Tip   In production environments, make sure you set the logging level to a sparser level such as WARN or ERROR because voluminous logging output will hurt performance.

Configuring Automatic Loading of Configuration Files

You configure Logback to scan for changes in the configuration file and automatically reconfigure itself when you make any logging-related configuration changes. You do this by setting the scan attribute of the <configuration> element to true:

```
<configuration scan="true">
  ...
</configuration>
```

Configuring Appenders

You configure appenders with the <appender> element. This element takes two mandatory attributes: name and class. The name attribute names the appender, and the class attribute points to the name of the appender class.

The <appender> element may additionally contain <layout>, <encoder>, and <filter> elements.

Setting Up Log Rotation

By default, Cassandra rolls the system.log file once it reaches 20MB in size. It also compresses the older archived log files in zip format and names them as system. log.1.zip, system.log.2.zip, and so on.

You can configure the following rolling policies in Logback:

- Location and name of the log file

- Location and name of the archive file

- Maximum log file size to trigger rolling to a new log file

- The log levels

The default logging level is INFO. You can set ALL, TRACE, DEBUG, WARN, and ERROR as the logging level. You can turn off all logging by setting the OFF logging level.

Using Nodetool to Set the Logging Level for a Service

Instead of editing the logback.xml file, you can run the nodetool setlogginglevel command to set the logging level from the command line.

Setting the Log Level

The following is the syntax for setting the log level of a service from the command line with the nodetool setlogginglevel command:

```
$ nodetool <options> setlogginglevel -- < class_qualifier > < level >
```

In this command,

- class_qualifier is the logger class qualifier, for example, org. apache.cassandra.service.StorageProxy.

- level is the logging level.

You can set the following values for the class_qualifier option:

- org.apache.cassandra

- org.apache.cassandra.db

- org.apache.cassandra.service.StorageProxy

You can choose the log level from among ALL, TRACE, DEBUG, INFO, WARN, and ERROR. You can set the level attribute to OFF to turn off all logging.

You can see more details about the operations the database is performing by setting the logging level to DEBUG. The following example shows how to set the StorageProxy service to the DEBUG level:

```
$ nodetool setlogginglevel org.apache.cassandra.service.StorageProxy DEBUG
```

Checking the Current Log Level

You can check the current log level with the nodetool getlogginglevels command, as shown here:

```
$ nodetool getlogginglevels
Logger Name                                      Log Level
ROOT                                                INFO
com.thinkaurelius.thrift                            ERROR
org.apache.cassandra                                DEBUG
org.apache.cassandra.service.StorageProxy           INFO
$
```

Monitoring Cassandra with Nagios

In a production environment, it's good to use a powerful enterprise-wide monitoring tool to check on your Casandra cluster. Nagios is a popular open source monitoring system, and it offers numerous capabilities to monitor Cassandra clusters.

Nagios uses the Nagios Remote Plugin Executor (NRPE) to monitor remote hosts on which Cassandra instances are running.

You need to install and configure the NRPE agents on the remote hosts. NRPE requires Nagios plugins, so you need to install the plugins as agents to monitor the local resources on the remote hosts.

In the following sections, I show how to install Nagios 4 on an Ubuntu server. I then show how to install NRPE and configure it. I start with the prerequisites you need to work with Nagios.

Installing the LAMP Stack and Addressing Other Prerequisites

A LAMP stack is a set of open source software that enables servers to host dynamic websites and web applications. The LAMP acronym stands for the following four components:

- The Linux operating system

- The Apache web server

- The MySQL database

- PHP

In the following sections I explain how to install LAMP. You likely already have the Linux OS, so you need to install the Apache web server, the MySQL database, and PHP.

1. Install and start the Apache web server.

```
$ sudo apt-get install apache2
$ sudo systemctl restart apache2
```

2. Install the MySQL database.

```
$ sudo apt-get install mysql-server
```

You can run the MySQL security script at this point to remove insecure default configurations and lock down database access.

```
$ mysql-secure_installation
```

3. Install PHP. The following command will install PHP 5:

```
$ sudo apt-get install php libapache2-mod-php php-mcrypt
php-mysql
```

4. By default, when a server requests a directory, Apache will first look for a file named index.html. You want to make Apache look first for an index.php file instead. To do this, edit the /etc/apache2/mode-enabled/dir.conf file, which looks as follows:

```
<IfModule mod_dir.c>
    DirectoryIndex index.html index.cgi index.pl index.php
index.xhtml index.htm
</IfModule>
```

Just move the index.php file to the head of the line, right before the DirectoryIndex file. Close and save the file.

5. Restart the Apache server so the changes you made in Step 4 go into effect.

```
$ sudo systemctl restart apache2
```

6. Check that the MySQL and PHP are running O.K. You can confirm the status of the Apache server by running the following command:

```
$ sudo systemctl status apache2
```

7. And you can confirm that you've properly configured PHP by running a basic PHP script named info.php (/var/www/html/info.php).

```
<?php
pPhpinfo();
?>
```

Once you do this, visit the following address in a browser:

```
http://<your -server_IP_address>/info.php
```

You should see a nice web page showing details about the PHP installation. You're good to go as far as the LAMP stack is concerned. Before you can install Nagios, however, you must complete several prerequisites, which I explain next.

Creating the Nagios User and Group

You need a user to run the Nagios server. Create a user named *nagios* that belongs to a group named *nagcmd*.

```
$ sudo useradd nagios
$ sudo groupadd nagcmd
$ sudo usermod -a -G nagcmd nagios
```

Installing the Build Dependencies

You're installing Nagios Core (the free version) from source, so you'll need to install additional binaries to complete the build. You'll also need apache2-utils to set up your Nagios web interface. You can install all the requisite packages thus:

```
$ sudo apt-get install build-essential libgd2-xpm-dev openssl libssl-dev
xinetd apache2-utils
```

Finally, with all the prerequisites completed, you're ready to install Nagios itself.

Installing Nagios

To install Nagios, you must download the source, configure it, and compile (build) it.

Downloading Nagios

Download the source for the latest stable Nagios version and extract the archive, as shown here:

```
$ cd ~
$ curl -L -O https://assets.nagios.com/downloads/nagioscore/releases/
nagios-4.1.1.tar.gz
$ tar xvzf nagios-*.tar.gz
$ cd nagios-*
```

Configuring Nagios

Before you can build Nagios Core, you must install the packages.

```
$ ./configure --with-nagios-group=nagios --with-command-group=nagcmd
```

Compiling and Installing Nagios

Compile Nagios and run additional commands to install Nagios.

```
$ sudo make
$ sudo make install
$ sudo make install-commandmode
$ sudo make install-init
$ sudo make install-config
$ sudo /usr/bin/install -c -m 644 sample-config/httpd.conf /etc/apache2/
sites-available/nagios.conf
```

Finally, add the web server user *www-data* to the *nagcmd* group, so you can issue external commands to Nagios via the web interface.

```
$ sudo usermod -G nagcmd www-data
```

Installing the Nagios Plugins

The next big step is to install the Nagios plugins. You follow similar steps as you did with the Nagios server installation in the previous section. I summarize the steps here.

```
$ cd ~
$ curl -L -O http://nagios-plugins.org/download/nagios-plugins-2.1.1.tar.gz
$ tar xvf nagios-plugins-*.tar.gz
$ cd nagios-plugins-*

$ ./configure --with-nagios-user=nagios --with-nagios-group=nagios --with-
openssl

$ make
$ sudo make install
```

You can verify the configuration thus:

```
$ sudo /usr/local/nagios/bin/nagios -v /usr/local/nagios/etc/nagios.cfg
...
Running pre-flight check on configuration data...
...
$
```

Note   You can easily create custom plugins for Nagios. The only requirements are that the plugin be executable through the command prompt, and that it return one of these exit values: O.K., Warning, Critical, or Unknown state.

Installing NRPE

The downloading, configuring, and building of NRPE follows the same steps as that of Nagios and the Nagios plugins. The following are the steps:

```
$ cd ~
$ curl -L -O http://downloads.sourceforge.net/project/nagios/nrpe-2.x/
nrpe-2.15/nrpe-2.15.tar.gz
$ tar xvf nrpe-*.tar.gz
$ cd nrpe-*
$ ./configure --enable-command-args --with-nagios-user=nagios --with-
nagios-group=nagios --with-ssl=/usr/bin/openssl --with-ssl-lib=/usr/lib/
x86_64-linux-gnu

$ make all
$ sudo make install
$ sudo make install-xinetd
$ sudo make install-daemon-config
```

Add the private IP address of the Nagios server at the end of the only_from line in the *xinetd* startup script, /etc/xinetd.d/nrpe.

```
only_from = 127.0.0.1 192.168.159.129
```

Restart the xinetd service to start NRPE.

```
$ sudo service xinetd restart
```

Configuring Nagios and Apache

Although you installed and built Nagios earlier, you still have some more configuring left to do for both the Nagios server and the Apache web server.

Configuring Nagios

Edit the main Nagios configuration file (/usr/local/Nagios/etc/nagios/cfg) as follows:

```
cfg_dir=/usr/local/nagios/etc/servers
```

Each of the hosts in your Cassandra cluster will have a configuration file that's stored in this directory. You must create the directory.

```
$ sudo mkdir /usr/local/nagios/etc/servers
```

It's also a good idea to edit the contacts configuration file (/usr/local/nagios/etc/objects/contacts.cfg), and replace the value for *email* with your own email address.

You also need a new command at the end of the Nagios configuration file, /usr/local/nagios/etc/objects/commands.cfg.

```
define command{
        command_name check_nrpe
        command_line $USER1$/check_nrpe -H $HOSTADDRESS$ -c $ARG1$
}
```

Configuring Apache

You must next configure the Apache web server, as explained in the following steps.

1. Enable the Apache rewrite and cgi modules, as shown here:

   ```
   $ sudo a2enmod rewrite
   $ sudo a2enmod cgi
   ```

2. Create an admin user named *nagiosadmin* using the *htpasswd*
    utility, which will help you access the Nagios web interface.

    ```
    $ sudo htpasswd -c /usr/local/nagios/etc/htpasswd.users
    nagiosadmin
    New password:
    Re-type new password:
    Adding password for user nagiosadmin
    ```

3. Start the Nagios and Apache servers.

    ```
    $ sudo service nagios start
    $ sudo service apache2 restart
    ```

Adding the Cassandra Cluster Hosts for Monitoring

There are three things you need to do to enable the monitoring of your Cassandra hosts
via Nagios:

- Install and configure NRPE on the cluster's hosts (nodes).

- Add the hosts to the Nagios server configuration.

- Add the services you want to monitor on the hosts.

Installing and Configuring NRPE on the Hosts

On each of the hosts that you'd like to monitor, do the following.

1. Update apt-get.

    ```
    $ sudo apt-get update
    ```

2. Install the Nagios plugins and the Nagios remote plugin executor
    (NRPE).

    ```
    $ sudo apt-get install nagios-plugins nagios-nrpe-server
    ```

3. Update the NRPE configuration file (/etc/nagios/nrpe.cf)
    and add the IP address of the Nagios server as a value for the
    allowed_hosts property.

    ```
    allowed_hosts=127.0.0.1,172.31.22.133
    ```

4. Restart NRPE.

```
$ sudo service nagios-nrpe-server restart
```

Adding Hosts to the Nagios Server Configuration

Create a separate configuration file on the Nagios server for each of the Cassandra hosts you want to monitor. For example, for a server named *host1*, you must create the file host1.cfg (/usr/local/nagios/etc/servers/host1.cfg).

In the host1.cfg file, add the host definition, as shown here:

```
define host {
        use                             linux-server
        host_name                       host1
        alias                           Nagios Agent 1
        address                         192.168.159.130
        max_check_attempts              5
        check_period                    24x7
        notification_interval           30
        notification_period             24x7
}
$ sudo service nagios start
```

At this time, Nagios can monitor just two basic things: it can tell you if the host is up or down. You need to configure the services you want Nagios to monitor by adding the services, as shown in the next section.

Adding Cassandra-Specific Plugins

The next step is to show you how to download and test a Cassandra-specific plugin. This plugin is named *Check Cassandra Status and Heap Memory* and it can report on a node's status (UP/DOWN) as well as memory heap utilization WARNING/CRITICAL notifications.

Follow these steps to work with the plugin.

1. Change the working directory to the /user/local/nagios/
   libexec directory.

   ```
   $ cd /usr/local/nagios/libexec
   ```

2. Download the plugin (cassandra.pl) and change permissions
   to it.

   ```
   $ wget "https://exchange.nagios.org/components/com_mtree/
   attachment.php?link_id=3819&cf_id=24"
   -O cassandra.pl
   $ chmod +x cassandra.pl
   $ chown nagios:nagios cassandra.pl
   ```

3. Test to make sure the plugin works.

   ```
   $./cassandra.pl
   CASSANDRA OK -   | heap_mem=5.59
   $
   ```

   The output of the test shows that the plugin works fine. If you
   see an error message instead, edit the cassandra.pl file and
   make sure that the $nodetool_path variable points to the correct
   location for the nodetool binary.

4. Configure NREP so Nagios can execute the plugin. Edit the /usr/
   local/nagios/etc/nrpe.cfg file in the Cassandra server and add
   the following line at the end of the file:

   ```
   . command[cassandra]=/usr/local/nagios/libexec/cassandra.
   pl $ARG1$
   ```

5. Restart the xinetd service on the Cassandra server.

   ```
   $ sudo service xinetd restart
   ```

6. Open a session with the server on which Nagios is running, and
   execute the following command:

```
$ sudo /usr/local/nagios/libexec/check_nrpe -H
192.168.159.130 -c cassandra
CASSANDRA OK -  | heap_mem=10.32
$
```

The output shows you've correctly configured NRPE on the Cassandra server. You can then add this configuration to the Nagios server.

Summary

Nodetool commands such as nodetool tablestats and nodetool tablehistograms help you learn key facts about the SSTables. Other nodetool commands such as nodetool status and nodetool info help monitor the cluster status.

I discussed logging at some length. Logging isn't a mere incidental byproduct of activity in a database. Logging is a valuable source of information for both what went right and what went wrong in a cluster. Cassandra allows you to configure logging in multiple ways, and customizing logging to suit your needs is a good strategy that helps you to effectively troubleshoot and monitor your databases.

JConsole is simple to set up and is helpful in tuning the performance of a database. I discussed Nagios at length, but there's a lot more to it, so I barely scratched the surface here. I used the free version of Nagios in my examples; the paid version offers more capabilities plus support! Whether it be Nagios or some other tool, a powerful graphical tool is essential to cluster management.

CHAPTER 11

Tuning Cassandra Performance

Cassandra has numerous configuration knobs, and nowhere is this truer than in the area of performance tuning, where the database administrator can control several aspects of performance. Out of the box, Cassandra runs with default settings for many performance-related configuration properties. As an administrator, you should know these defaults and modify them to suit your environment.

This chapter focuses on the following areas of Cassandra performance:

- Tracing queries to analyze database performance

- Caching data

- Compaction strategies

- Compression

- Tuning Bloom filters

- Tuning the JVM and garbage collection strategies

- Stress testing with the `cassandra-stress` tool

Using Tracing to Analyze Performance

When you're tuning database performance, you can turn on tracing in a database to get detailed transaction information about the internal operations Cassandra performs. Tracing is valuable in tracking the time Cassandra spends during various stages of query

© Sam R. Alapati 2018
S. R. Alapati, *Expert Apache Cassandra Administration*, https://doi.org/10.1007/978-1-4842-3126-5_11

processing, such as the parsing and executing of a query. Tracing provides detailed timing information pertaining to actions such as the following:

- Preparing a SQL statement

- Read repair related activity

- Memtable and SSTable data lookups

- Key cache searches

- Interactions among a cluster's nodes

Tracing data is highly useful in evaluating query efficacy. For example, inappropriate or superfluous secondary indexes show up in a high inter-node activity among the nodes when you trace a query that involves those indexes.

Note   Turning tracing on will help trace two broad types of activity in a cluster: queries and repair operations.

By default, Cassandra disables tracing. Once you enable tracing, Cassandra captures the transaction details in the system_traces keyspace. Two tables in this keyspace hold the trace data:

- system_traces.session: Stores high-level details of transactions, such the length of the transaction and the session ID.

- system_traces.events: Stores detailed information about all operations the database performs.

Note   Cassandra can only store the tracing data for a limited period. So, if you wish to hold on to tracing data for a longer period, you must configure the following properties:

tracetype_query_ttl: Sets the TTL for different trace types used during logging of the query process. Default value is 86,400 seconds (1 day).

tracetype_repair_ttl: Sets the TTL for different trace types used during logging of the repair process. Default value is 604,800 seconds (about a week).

The types of database activity that the system_traces.events table captures reveal the operations that Cassandra performs when you issue a query.

```
cqlsh> select  activity from system_traces.events;

 activity
-----------                       Parsing select id from cycling.cyclist_name;
Preparing statement
Computing ranges to query
 Submitting range requests on 513 ranges with a concurrency of 112 (0.9
 rows per range expected)
 Executing seq scan across 0 sstables for (min(-9223372036854775808),
 max(-9196210656004250337)]
 Read 0 live and 0 tombstone cells
 Enqueuing request to /192.168.159.129
 Sending RANGE_SLICE message to /192.168.159.129
RANGE_SLICE message received from /192.168.159.130
Enueuing request to /192.168.159.129
Enqueuing response to /192.168.159.130
Sending RANGE_SLICE message to /192.168.159.129
Processing response from /192.168.159.129
RANGE_SLICE message received from /192.168.159.130
Executing seq scan across 0 sstables for (max(-8175039996930460291),
max(-8066091505323311933)]
 Read 0 live and 0 tombstone cells
 ...
cqlsh>
```

Managing Tracing

You can check the status of tracing thus:

```
cqlsh> tracing
Tracing is currently disabled. Use TRACING ON to enable.
cqlsh>
```

It's simple to turn on tracing in the database; just run the tracing on command.

```
cqlsh> tracing on;
Now Tracing is enabled
cqlsh> tracing;
Tracing is currently enabled. Use TRACING OFF to disable
cqlsh>
```

To turn off all tracing, run the tracing off command.

```
cqlsh> tracing off;
Disabled Tracing.
cqlsh>
```

Managing Probabilistic Tracing

Cassandra uses a probabilistic tracing strategy wherein you can configure the probability of the percentage of statements that the database will trace. The probability is on a per-node basis. A probability of 1.0 means that the database will trace all SQL statements, and a setting of 0.5 means it's likely to trace about 50 percent of all requests.

The default probability is 0.0, which means tracing is disabled. Higher probability settings involve a higher number of writes, and sometimes this can adversely affect cluster performance, so you must exercise caution when setting the tracing probability.

Cassandra recommends that you start with a small probability setting such as 0.001 and raise it gradually based on what's happening in the database. Even a low setting such as this can have a significant impact on database performance, so be sure to look out for a performance hit.

The nodetool gettraceprobability command shows the current trace probability.

```
$ nodetool gettraceprobability
Current trace probability: 0.0
$
```

You can set the trace probability by running the nodetool settraceprobability command for tracing a read or write request, as shown here:

```
$ nodetool settraceprobability 0.1
$ nodetool gettraceprobability
Current trace probability: 0.1
$
```

The system_traces.sessions table shows valuable information about the traces.

```
cqlsh> select * from system_traces.sessions;
 session_id     | client   | command | coordinator   | duration |
 parameters     | request       | started_at
----------------+ ---------+---------+---------------+----------+
----------------d7103660-81e4-11e7-a4ad-89801d899afb
| 127.0.0.1 |   QUERY | 192.168.159.130 |   506051 |
{'consistency_level': 'ONE', 'page_size': '100', 'query': 'select * from
cycling.cyclist_name;', 'serial_consistency_level': 'SERIAL'} | Execute
CQL3 query | 2017-08-15 18:09:11.878000+0000
   189ff700-81e5-11e7-a4ad-89801d899afb | 127.0.0.1 |   QUERY |
192.168.159.130 |     6848 | {'consistency_level': 'ONE', 'page_size':
'5000', 'query': 'SELECT * FROM system_traces.sessions WHERE session_id
= 185a6280-81e5-11e7-a4ad-89801d899afb', 'serial_consistency_level':
'SERIAL'} | Execute CQL3 query | 2017-08-15 18:11:01.872000+0000
185a6280-81e5-11e7-a4ad-89801d899afb | 127.0.0.1 |   QUERY |
192.168.159.130 |   177343 |                    {'consistency_level':
'ONE', 'page_size': '100', 'query': 'select * from cycling.cycle;',
'serial_consistency_level': 'SERIAL'} | Execute CQL3 query | 2017-08-15
18:11:01.416000+0000
...
(11 rows)
cqlsh>
```

How to Trace Write and Read Requests

Tracing is resource intensive and it can consume vast amounts of storage. Tracing selectively is the right strategy, instead of enabling tracing for the entire database. In this section, I show how to trace specific read and write requests.

Tracing a Write Request

Here's a simple example that shows how to get trace information for a write request (insert statement). You first turn on tracing and then insert a row in the cyclist_name table (output shows partial data).

```
cqlsh> tracing on;
Now Tracing is enabled
cqlsh> insert into cycling.cyclist_name ( id, lastname, firstname ) values
(uuid(), 'FRAME', 'Nina' );

Tracing session: bdfdf410-8116-11e7-81bd-63c57c069fd1
 activity                                                                |
timestamp                      | source            | source_elapsed | client
Execute CQL3 query | 2017-08-14 10:33:53.489000 | 192.168.159.130 |
              0 | 127.0.0.1
Determining replicas for mutation [Native-Transport-Requests-1] | 2017-08-
14 10:33:53.522000 | 192.168.159.130 |               32844 | 127.0.0.1
MUTATION message received from /192.168.159.130 [MessagingService-
Incoming-/192.168.159.130] | 2017-08-14 10:33:53.530000 | 192.168.159.129 |
          843 | 127.0.0.1
Sending MUTATION message to /192.168.159.129 [MessagingService-Outgoing-/
192.168.159.129-Small] | 2017-08-14 10:33:53.532000 | 192.168.159.130
|          42754 | 127.0.0.1
Appending to commitlog [MutationStage-1] | 2017-08-14 10:33:53.532000 |
192.168.159.129 |               2668 | 127.0.0.1
Adding to cyclist_name memtable [MutationStage-1] | 2017-08-14
10:33:53.532000 | 192.168.159.129 |               3181 | 127.0.0.1
Processing response from /192.168.159.129 [RequestResponseStage-3] |
2017-08-14 10:33:53.545000 | 192.168.159.130 |               56321 | 127.0.0.1
```

```
      Request complete | 2017-08-14 10:33:53.546583 |
192.168.159.130 |            57583 | 127.0.0.1
```

The trace data shows the following bits of information:

- The target nodes where Cassandra replicates the row you've inserted

- How Cassandra appends the new data to the commitlog

- How the database adds the new data to the memtable

- How the database confirms ("Request complete") that the request to insert the data was successfully completed

Tracing a Read Request

Tracing a read request produces a lot more data than tracing a write request, so be careful when you trace these requests. Part of the reason for the dense trace data is because Cassandra spreads rows across multiple SSTables, and so it must read multiple SSTables to retrieve data. The trace shows all requests that the database makes to fulfill the read request.

Here's a partial output for a simple select statement:

```
cqlsh> select * from cycling.cyclist_name;
```

```
 id                                   | firstname | lastname
--------------------------------------+-----------+-----------------
 69f1cb04-687d-4be7-a91f-72ef037c5514 |     Sammy |           FRAME
 e7ae5cf3-d358-4d99-b900-85902fda9bb0 |      Alex |           FRAME
 d378af39-1a28-474c-836b-aa960fed6f2b |      Nina |           FRAME
 fb372533-eb95-4bb4-8685-6ef61e994caa |   Michael |        MATTHEWS
 5b6962dd-3f90-4c93-8f61-eabfa4a803e2 |  Marianne |             VOS
 220844bf-4860-49d6-9a4b-6b5d3a79cbfb |     Paolo |       TIRALONGO
 6ab09bec-e68e-48d9-a5f8-97e6fb4c9b47 |    Steven |      KRUIKSWIJK
 e7cd5752-bc0d-4157-a80f-7523add8dbcd |      Anna | VAN DER BREGGEN

(8 rows)
```

```
Tracing session: b2b0eda0-8117-11e7-81bd-63c57c069fd1
 activity                                             |
timestamp                      | source        | source_elapsed | client
Execute CQL3 query | 2017-08-14 10:40:44.026000 | 192.168.159.130
|          0 | 127.0.0.1
Parsing select * from cycling.cyclist_name; [Native-Transport-Requests-1] |
2017-08-14 10:40:44.027000 | 192.168.159.130 |            364 | 127.0.0.1
Preparing statement [Native-Transport-Requests-1] | 2017-08-14
10:40:44.027000 | 192.168.159.130 |            503 | 127.0.0.1
RANGE_SLICE message received from /192.168.159.130 [MessagingService-
Incoming-/192.168.159.130] | 2017-08-14 10:40:44.073000 | 192.168.159.129
|          22 | 127.0.0.1
Sending REQUEST_RESPONSE message to /192.168.159.130 [MessagingService-
Outgoing-/192.168.159.130-Small] | 2017-08-14 10:40:44.086000 |
192.168.159.129 |          13202 | 127.0.0.1
RANGE_SLICE message received from /192.168.159.130 [MessagingService-
Incoming-/192.168.159.130] | 2017-08-14 10:40:44.095000 | 192.168.159.129
|          18 | 127.0.0.1
REQUEST_RESPONSE message received from /192.168.159.129 [MessagingService-
Incoming
...                        Processing response from /192.168.159.129
[RequestResponseStage-5] | 2017-08-14 10:40:44.557001 | 192.168.159.130
|          531160 | 127.0.0.1
Request complete | 2017-08-14 10:40:44.562582 | 192.168.159.130
|          536582 | 127.0.0.1
cqlsh>
```

The output shows how Cassandra parses the CQL statement and prepares it before sending a request message, to which the receiving node responds with a response message. Finally, the Request Complete comment shows that the request was successfully processed by the node(s).

Tuning Bloom Filters

As I explained in Chapter 5, Bloom filters are a performance aid for Cassandra. They help the database during an index scan by letting it know if an SSTable has data for a specific partition. When clients request data, the Bloom filter checks if the row exists before the database performs disk I/O.

Configuring Bloom filters involves a tradeoff between memory usage and the probability of finding the data quickly, using less I/O than otherwise.

Configuring Bloom Filters

You can set the `bloom_filter_fp_chance` attribute for a table to a value between 0 and 1. As you go from 0 to 1, you use less memory. A value of 0 means you set the largest value for the Bloom filter and use the highest amount of memory. Setting it to 1 means that you've disabled Bloom filters.

The default value for the `bloom_filter_fp_chance` attribute depends on the compaction strategy in place. The following are the values for this attribute under Cassandra's three main compaction strategies:

- For the LeveledCompactionStrategy: 0.1

- For the SizeTieredCompactionStrategy, DataTieredCompactionStrategy, and TimeWindowCompactionStrategy: 0.01.

The recommended value for the `bloom_filter_fp_chance` attribute is 0.1. Higher values aren't necessarily helpful since raising the value produces diminishing returns. This is because higher values consume more memory but yield a disproportionately smaller performance benefit. You can set the property at the time you create a table, or later.

Regenerating Bloom Filters

Whenever you change the value of the `bloom_filter_fp_chance` attribute for a table, you must regenerate the Bloom filters. You can regenerate the Bloom filters in two ways: you can run a manual compaction of the table or upgrade the SSTables.

To manually compact a table, run the `nodetool compact` command. The `nodetool compact` command has the following syntax:

```
nodetool [options] compact [(-et <end_token> | --end-token <end_token>)]
[(-s | --split-output)] [(-st <start_token> | --start-token <start_token>)]
[--] [<keyspace> [<tables>...]]
[--user-defined] <SSTable file>...
```

You must specify the compaction strategy by selecting from the STCS, TWCS, and the LCS compaction strategies. By default, the command runs major compaction on all keyspaces and tables in the database, but you can limit the compaction to one or more tables.

Manually compacting SSTables to regenerate Bloom filters is almost never a good strategy. You can instead upgrade an SSTable when you change the value of the `bloom_filter_fp_chance` property for that table. You do this by running the `nodetool upgradetsstables` command. The command has the following syntax:

```
$ nodetool <options> upgradesstables
 ( -a | --include-all-sstables )
 -- <keyspace>  <table> ...
```

Caching Data

To optimize Cassandra's use of cache memory, you can configure the caching property for a table. Caching helps Cassandra perform a warm restart of a node, wherein the database periodically stores the cache to disk and reads it back into cache when you restart the node. Not having the cache means the node will take longer to restart.

Caching is useful for very busy clusters. Caching data is not advisable for data that's not in high demand. Therefore, it's a good idea to separate heavily read data into separate tables so you can cache the data in those tables. It also isn't a good idea to cache data when dealing with tables that contain extremely long partitions.

Types of Data Caches

There are two main types of data caches: the key cache and the row cache. (In addition, you can also require Cassandra to cache *counters*, as I explain later.)

- *Key cache*: The key cache, also called the partition key cache, caches the partition index for an SSTable. The database requires less disk reads to fetch data from the key cache instead of getting the data either from disk or from the OS page cache. Since the key cache offers more bang for the buck in the sense that it increases the possibility of cache reads without requiring too much memory, Cassandra enables it by default.

- *Row cache*: You can ask Cassandra to cache a specific number of rows in a partition. You set the number of rows to cache with the rows_per_partition table option. Since a row cache stores entire rows instead of just the keys, it could lead to performance deterioration when dealing with large data sets, which the database may have to read from disk since memory may not be sufficient to store all the data. Configure the row cache carefully since it may sometimes hurt performance rather than help it.

Here are the key principles you ought to remember regarding the use of caching in a Cassandra database:

- You don't want to set both types of caching for a table. Pick either partition key caching or row caching for a table.

- Row caching is more space intensive than key caching since it stores entire rows. Therefore, it's a good idea to use row caching selectively, using it to cache only those rows that are frequently accessed by users and clients.

- As to which tables you must cache, read-heavy tables are a good target for caching. When the reads overwhelmingly outweigh writes for a table, caching is something you want to consider.

Note   Your users won't be reading archived tables often so you can disable caching for those tables.

Where Cassandra Stores the Cached Data

Cassandra stores the row and key caches in a directory you can specify with the configuration property `saved_caches_directory`. The default location of the `saved_caches_directory` property is determined as follows:

- `/var/lib/cassandra/saved_caches`   `/* for package installations`

- `install_location/data/saved_caches`   `/* for tarball installations`

Configuring Caching

You can set two attributes that determine caching behavior:

- `keys`: This property can take the values `ALL` or `NONE`. The value `ALL` means all primary keys or rows. The value `NONE` means no primary keys or rows. The default value is `ALL`.

- `rows_per_partition`: This property helps you set the number of rows the database must cache in a partition. This attribute has three values: `ALL, NONE,` and `N`. The values `ALL` and `NONE` have the same meaning as for the `keys` attribute. The value `N` specifies the number of rows per partition. The default value is `NONE`.

You set the table caching property to configure the partition key cache and the row cache, either when creating a table or later. The following is the syntax for the `caching` table property:

```
caching = {
 'keys' = 'ALL | NONE',
 'rows_per_partition' = 'ALL' | 'NONE' |N}
```

By default, this is what Cassandra uses for caching data:

```
{ 'keys' : 'ALL', 'rows_per_partition' : 'NONE' }
```

The following are a few examples that show how to configure caching properties. The first example shows how to configure caching for a table. You specified NONE as the value for the keys attribute so no primary keys or rows are cached. The database will cache 120 rows per partition.

```
CREATE TABLE test (
  userid text PRIMARY KEY,
  first_name text,
  last_name text,
)
WITH caching = { 'keys' : 'NONE', 'rows_per_partition' : '120' };
```

You can cache materialized views in addition to SSTables. The following example shows how to specify caching when creating a materialized view:

```
CREATE MATERIALIZED VIEW cycling.cyclist_by_age
AS SELECT age, name, country
FROM cycling.cyclist_mv
WHERE age IS NOT NULL AND cid IS NOT NULL
PRIMARY KEY (age, cid)
WITH caching = { 'keys' : 'ALL', 'rows_per_partition' : '100' }
;
```

The following example shows how to run the ALTER TABLE statement to cache all cyclists in every age partition:

```
ALTER MATERIALIZED VIEW cycling.cyclist_by_age
WITH caching = {
 'keys' : 'ALL',
 'rows_per_partition' : 'ALL' } ;
```

Global Caching Parameters

In the previous section, I explained how you can configure caching by setting caching properties for a table. Cassandra also offers several *global caching properties,* which you can configure in the cassandra.yaml file. I explain the global caching parameters in the following sections.

Configuring the Size of the Row Cache

When you configure row caching, you must set the number of rows per partition. There's no fixed rule stating that X percentage of rows is the right amount of rows to cache. Instead, you size the row cache based on the workload patterns in your database.

Note   By default, row caching is disabled and key caching is enabled.

You configure the size of the row cache at the database level by setting a value for the row_cache_size_in_mb property in the cassandra.yaml file. This parameter determines the maximum amount of memory Cassandra can allocate to store rows from the most frequently read table partitions of a table. By default, the database doesn't cache rows so the parameter's default value is 0.

A row cache potentially saves more time than the key cache, but it requires more space since it caches the entire row. Therefore, you must use the row cache just for rows in heavy demand or for rows that don't change. Too low a value for the row_cache_size_in_mb property may result in Cassandra not loading some of the hot keys when the database starts up.

Configuring the Size of the Key Cache

The key_cache_size_in_mb parameter allows you to set the maximum size of the key cache for all tables in a database. The value of this parameter isn't set by default, and Cassandra sets the key cache to 5% of the heap size or 100MB, whichever is smaller. You can disable key caching by setting the key_cache_size_in_mb parameter to 0.

Configuring the Frequency of Caching

You can configure how frequently the database caches rows caches and partition key caches to disk.

- The row_cache_save_period property determines how long (in seconds) the database retains the rows in the cache before saving them to the directory specified by the saved_caches_directory property. Row caching is disabled by default, meaning the default value of the row_cache_save_period parameter is 0.

- The key_cache_save_period parameter determines how long the database keeps keys in the cache before saving them to the directory specified by the saved_caches_directory property. The default value is 4 hours (14,400 seconds).

Specifying the Number of Keys to Save

By default, the database saves all keys from the row cache. You can configure the row_cache_keys_to_save parameter to specify the number of keys to save. The default value for this parameter is disabled, which means that all keys are saved.

Since by default the database doesn't enable the row cache, the default value for the row_cache_class_name parameter is disabled. You can specify one of the following two values as the class name for the row cache provider to use:

- OHCProvider: Fully off-heap

- SerializingCacheProvider: Partially off-heap

The OHCProvider is newer, and benchmarking tests show that it offers roughly 15 percent better performance than the older, partially off-heap row cache provider.

Using the Counter Cache

Besides the row and key caches, Cassandra also lets you configure a counter cache. The counter cache stores counters that are in high demand, thus reducing the contention for the counter cells. When the database finds a counter in the cache (cache hit), it needs to hold the counter locks for a shorter period, speeding up the updating of the hot counter cells.

Cassandra enables the counter cache by default. The size of the counter cache depends on the value you configure for the counter_cache_size_in_mb parameter. By default this parameter isn't set, and Cassandra uses the smaller of two values: 2.5% of the heap or 50MB.

As with the row and key caches, Cassandra saves the counter cache (just the keys) to disk in the directory you specify with the saved_caches_directory parameter. By default, this parameter has the value 2 hours (7,200 seconds).

The other configuration property for managing the counter cache is the counter_cache_keys_to_save parameter. You configure this property to tell the database how many keys from the counter cache it must save. By default this property is disabled, meaning Cassandra saves all keys.

If you perform counter deletes and use a low gc_grace_seconds setting, you should disable the counter cache, which you can do by setting the value of the counter_cache_size_in_mb parameter to zero.

Monitoring Caching

The nodetool info command helps you monitor caching performance, thus helping you adjust the row cache and key cache configuration based on current cache behavior. Here's an example:

```
$ nodetool info
ID                       : 0dbb9e0e-867e-4179-b6b6-631d38dd68f9
Gossip active            : true
Thrift active            : false
Native Transport active: true
Load                     : 12.46 MiB
Generation No            : 1503673044
Uptime (seconds)         : 91
Heap Memory (MB)         : 117.82 / 1492.00
Off Heap Memory (MB)     : 0.07
Data Center              : datacenter1
Rack                     : rack1
Exceptions               : 0
Key Cache                : entries 47, size 3.94 KiB, capacity 74 MiB, 69
hits, 120 requests, 0.575 recent hit rate, 14400 save period in seconds
Row Cache                : entries 0, size 0 bytes, capacity 0 bytes, 0 hits,
0 requests, NaN recent hit rate, 0 save period in seconds
Counter Cache            : entries 0, size 0 bytes, capacity 37 MiB, 0 hits,
0 requests, NaN recent hit rate, 7200 save period in seconds
Chunk Cache              : entries 30, size 1.88 MiB, capacity 341 MiB, 44
misses, 182 requests, 0.758 recent hit rate, 1072.447 microseconds miss
latency
Percent Repaired         : 100.0%
Token                    : (invoke with -T/--tokens to see
$
```

The `nodetool info` command shows the following caching-related information for all three of Cassandra's caches (row, key, and counter):

- Number of entries in the cache

- Size of the cache in bytes

- Capacity of the cache

- Number of requests for cached data

- Number of cache hits

- The cache hit ratio (rate)

- The save period in seconds

Tracing Database Operations to Optimize Caching

You can trace database reads to check if the read operations are getting data from the cache or straight from the SSTables stored on disk. When you trace a query that refers to a cached table, initially the trace shows that there was a row cache miss. When you rerun the query, the trace reveals the "row cache hit" line, as shown here:

```
row cache miss [ReadStage:23]
row cache hit [ReadStage:35]
```

A row cache miss means that the database has read the data from disk. A row cache hit, on the other hand, means that the database has found the data in the cache, saving itself the trouble of going to disk to read the data. Remember that a cache read is always many times faster than a read from disk.

The reason you see the "cache miss" early on is because it takes some time for the database to cache data, especially if the table is large. Once the database puts the data in the cache, subsequent queries will use the cached data instead of going to disk, meaning the query is going to return the results quite fast. You'll therefore see the "cache hit" message in the traces.

Once data gets in the cache, it stays there and the database uses it for all queries that refer to the data. However, if you update the data, it invalidates the cache. Cassandra ignores the cached data under conditions such as the following:

- If the query requires data from the cache as well as from disk (that is, cached+uncached data).

- When the query requests large amounts of data, which leads to the exceeding of the global cache size limit.

- The query requests data that's not at the beginning of the partition.

If any of these situations occur, you'll see the following line from the trace of a query that tries to use the cached data:

```
Ignoring row cache as cached value could not satisfy query [ReadStage:89]
```

In all of these cases, the cache proves to be insufficient to handle the query, forcing Cassandra to perform expensive (high resource cost) disk reads instead of using the more efficient cache reads. When you notice that the database is ignoring the row cache, depending on the cause, you can do one of the following to make the database use the cached data:

- Increase the cache size.

- Create a new table and place frequently accessed rows at the beginning of the partition.

- Limit the output size for a query (with the LIMIT N option in a SELECT statement) to keep the retrieved rows from exceeding the configured cache size.

Stress Testing Cassandra with cassandra-stress

Cassandra offers the cassandra-stress utility for benchmarking and load testing your clusters. The tool is also useful in testing changes to key aspects of database configuration, such as selecting a different compaction strategy. You can test read, write, and mixed workloads.

Cassandra-stress is useful for optimizing your data model, testing how well the database scales, and determining production capacity.

Running cassandra-stress

You can find the cassandra-stress tool in the $CASSANDRA_HOME/tools/bin directory.

Issue the command cassandra-stress to get a quick idea about the tool's capabilities, as shown here:

```
$CASSANDRA_HOME/tools/bin$ cassandra-stress
No command specified
Usage:      cassandra-stress <command> [options]
Help usage: cassandra-stress help <command>

---Commands---
read             : Multiple concurrent reads - the cluster must first
be populated by a write test
write            : Multiple concurrent writes against the cluster
mixed            : Interleaving of any basic commands, with
configurable ratio and distribution - the cluster must first be populated
by a write test
counter_write    : Multiple concurrent updates of counters.
counter_read     : Multiple concurrent reads of counters. The cluster
must first be populated by a counterwrite test.
user             : Interleaving of user provided queries, with
configurable ratio and distribution
help             : Print help for a command or option
print            : Inspect the output of a distribution definition
legacy           : Legacy support mode
version          : Print the version of cassandra stress

---Options---
-pop             : Population distribution and intra-partition visit order
-insert          : Insert specific options relating to various methods
for batching and splitting partition updates
-col             : Column details such as size and count distribution,
data generator, names, comparator and if super columns should be used
-rate            : Thread count, rate limit or automatic mode (default
is auto)
-mode            : Thrift or CQL with options
-errors          : How to handle errors when encountered during stress
-schema          : Replication settings, compression, compaction, etc.
-node            : Nodes to connect to
```

```
-log                  : Where to log progress to, and the interval at which
to do it
-transport            : Custom transport factories
-port                 : The port to connect to cassandra nodes on
-sendto               : Specify a stress server to send this command to
-graph                : Graph recorded metrics
-tokenrange           : Token range settings
$CASSANDRA_HOME/tools//bin$
```

At a high level, here are the key options you must be aware of:

- read: Multiple concurrent reads (you must first run a write test to populate the test tables with data)

- write: Multiple concurrent writes

- counter_write: Multiple concurrent counter updates

- counter_reads: Multiple concurrent counter reads (you must first populate the test tables with a counter_write test)

A cassandra-stress Example

When you run the cassandra-stress tool for the first time, it creates a keyspace named *keyspace1*, and within that keyspace, a table named either standard1 or counter1, depending on the type of tests you run. Cassandra reuses the keyspace and table it creates for all subsequent runs.

Running a Write Test

Here's an example that shows how to run a write test (remember that you must run a write test first before you can run a read test):

```
$ cassandra-stress write
n=100000
cl=one
****************** Stress Settings ******************
Command:
  Type: write
  Count: 100,000
```

```
Consistency Level: ONE
...
Insert:
  Revisits: Uniform:  min=1,max=1000000
  Visits: Fixed:  key=1
Columns:
  Max Columns Per Key: 5
  Column Names: [C0, C1, C2, C3, C4]
  Comparator: AsciiType
  Timestamp: null
  Variable Column Count: false
  Slice: false
  Size Distribution: Fixed:  key=34
  Count Distribution: Fixed:  key=5
Errors:
  Ignore: false
  Tries: 10
...
Schema:
  Keyspace: keyspace1
  Replication Strategy: org.apache.cassandra.locator.SimpleStrategy
  Replication Strategy Pptions: {replication_factor=1}
  Table Compression: null
  Table Compaction Strategy: null
  Table Compaction Strategy Options: {}
...
Connected to cluster: Test Cluster, max pending requests per connection
128, max connections per host 8
Datatacenter: datacenter1; Host: localhost/127.0.0.1; Rack: rack1
Created keyspaces. Sleeping 1s for propagation.
Running WRITE with 200 threads for 100000 iteration
type       total ops,    op/s,   pk/s,   row/s,    mean,     med,      .95,
.99,    .999,    max,   time,   stderr, errors,  gc: #,  max ms,  sum
ms,  sdv ms,      mb
...
```

```
Results:
Op rate                     :    3,011 op/s  [WRITE: 3,011 op/s]
Partition rate              :    3,011 pk/s  [WRITE: 3,011 pk/s]
Row rate                    :    3,011 row/s [WRITE: 3,011 row/s]
Latency mean                :   64.9 ms [WRITE: 64.9 ms]
Latency median              :   39.1 ms [WRITE: 39.1 ms]
Latency 95th percentile     :  190.2 ms [WRITE: 190.2 ms]
Latency 99th percentile     :  537.4 ms [WRITE: 537.4 ms]
Latency 99.9th percentile : 1155.5 ms [WRITE: 1,155.5 ms]
Latency max                 : 1632.6 ms [WRITE: 1,632.6 ms]
Total partitions            :    100,000 [WRITE: 100,000]
Total errors                :          0 [WRITE: 0]
...
Total operation time        : 00:00:33
END
$
```

In this command, you specify the write option to denote that you want to insert
rows into the table that cassandra-stress creates for the stress test. Cassandra-stress
inserts one million rows (n=10000000). The option c1 sets the consistency level to ONE.
The default consistency level is LOCAL ONE, and you can set any of the other standard
Cassandra consistency levels.

The previous example didn't use authentication. To use authentication when
running the cassandra-stress tool, specify the -mode command (with the option
native) to specify a username and password, as shown in this example:

```
$ cassandra-stress
-mode native cql3
user=cassandra password=cassandra
cl=QUORUM
```

The data you insert with the write option of the cassandra-stress tool is by default
never truncated. You can truncate the table that the cassandra-stress tool creates by
running the cassandra-stress command with the -truncate option:

```
$ cassandra-stress write n=100000000 cl=QUORUM truncate=always -schema
keyspace=keyspace-rate threads=200 -log file=write_$NOW.log
```

If you're specifying the mode command, make sure that you specify the truncate option before the mode option, as otherwise the cassandra-stress tool ignores the truncate option.

Running a Read Test

Once you populate the database tables created by the cassandra-stress tool as shown in the write test example, you have data that the tool can read. Now you can run the read test, as shown here:

```
$ cassandra-stress read n=10000000 -rate threads=50 duration=5
```

In this example, the options stand for the following:

- -read: Performs multiple concurrent reads of the data that you populated earlier. The suboption n=10000000 specifies the number of rows to read.

- -rate: Helps you specify the thread count, rate limit, or the automatic mode, which is the default. In this case, you specify the suboption thread count for the stress test by specifying threads=50.

- duration: Specifies that the test must read rows for a specific number of minutes (5 in this example).

Running a Mixed Workload

The cassandra-stress command option mixed helps you run commands with configuration ratio and distribution. Before running this command, you must populate the cluster with a write test.

Here's an example that shows how to run a mixed workload:

```
$ cassandra-stress mixed
ratio\(write=1,read=3\)
n=100000
cl=ONE
-pop dist=UNIFORM\(1..1000000\)
-schema keyspace="keyspace1"
-mode native cql3
-rate threads\>=16 threads\<=256
-log file=~/mixed_autorate_50r50w_1M.log
```

Note the following:

- The ratio option sets the ratio of writes to reads.

- The -pop command option sets the population distribution and intra-partition visit order. In this example, you specify pop dist=UNIFORM\ (1..1000000\), which means that of the 100,000 operations (n=100000), the test must select the keys uniformly distributed between 1 and 1,000,000. You specify the -pop option when you specify more data per node than can fit in the RAM of that node.

Setting the Replication, Compaction, and Compression Options

Run the -schema command option to view the options you can set for replication, compaction, and compression. It also enables you to specify the keyspace to use for the stress test.

```
$ cassandra-stress help  -schema
Usage: -schema [replication(?)] [keyspace=?] [compaction(?)] [compression=?]
  replication([strategy=?][factor=?][<option 1..N>=?]): Define the
  replication strategy and any parameters
      strategy=? (default=org.apache.cassandra.locator.SimpleStrategy)
      The replication strategy to use
      factor=? (default=1)                     The number of replicas
  keyspace=? (default=keyspace1)               The keyspace name to use
  compaction([strategy=?][<option 1..N>=?]): Define the compaction strategy
  and any parameters
      strategy=?                               The compaction strategy to use
  compression=?                                Specify the compression to use for
                                               sstable, default:no compression
$
```

The following example shows how to change the replication strategy to NetworkTopologyStrategy for a node named *cass1*:

```
$ cassandra-stress write n=500000 no-warmup -node cass1 -schema "replication
(strategy=NetworkTopologyStrategy, cass1=2)"
```

Running a Stress Test on Multiple Nodes

All of the earlier examples ran on a single node. Sometimes a single node can't handle the workload for a stress test. You can specify multiple nodes with the $NODES variable that you specify with the -node command option. You specify the nodes in a comma-delimited list of IP addresses with the $NODES variable.

The following example shows how to run a stress test on two nodes named *cass1* and *cass2*:

```
$ cassandra-stress write n=1000000 cl=one -mode native cql3 -schema
keyspace="keyspace1" -pop seq=1..1000000 -log file=~/node1_load.log
-node $NODES
```

Running cassandra-stress with a YAML-Based Profile

You can use a YAML-based *profile file* when running the cassandra-stress tool. The profile helps you define various compaction strategies and the cache settings that the database will use during the stress tests. You can find the sample YAML files provided by Cassandra in the following locations:

- `/usr/share/docs/cassandra/examples`   `/* Package installations`
- `$CASSANDRA_HOME/tools/`   `/* Tarball installations`

A Cassandra-provided YAML file, such as the `cqlstress-example.yaml` file, contains various configurations for the stress tests under separate sections. These configuration properties include keyspace and table definitions, as well as the query definition. You can add secondary indexes and materialized views under the *extra-definitions* section. You can modify the sample profile files or create your own profile files.

You can specify the YAML file as the value for the profile option when you run the cassandra-stress command. The profile option points to the location of the YAML file that you use.

The following example shows how to run a stress test with the help of the cqlstress-example.yaml file:

```
$ cassandra-stress user profile=$CASSANDRA_HOME/tools/cqlstress-example.
yaml n=1000000 ops\(insert=3,read1=1\) no-warmup cl=QUORUM
```

You can specify the -graph option to create a graph for the stress tests you run. You can view the graphs with a web browser after the test completes. Here's an example:

```
$ cassandra-stress
user profile=tools/cqlstress-example.yaml
ops\(insert=1\)
-graph file=test.html
title=test
revision=test1
```

You can view a graph of the stress test by viewing the file test1.html in a web browser. You must provide the name for the HTML file that captures the output of the stress test for viewing via a browser. The title and revision options aren't required, but if you're running multiple stress tests using the same data, you must use the revision parameter.

Figure 11-2 shows the graph of the stress test with the previous cassandra-stress command.

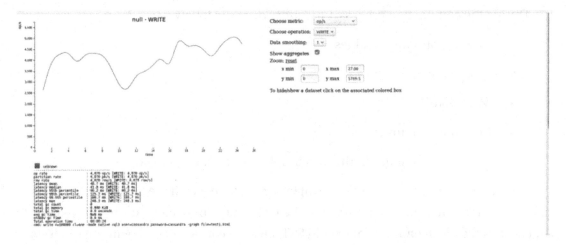

Figure 11-2. *A graph of a cassandra-stress tool stress test, generated with the -graph option*

Configuring Compaction Strategies

Compacting data helps consolidate SSTables, merge keys, combine columns, and evict tombstones.

As you learned in Chapter 6, Cassandra's SSTables are immutable, meaning that the database never overwrites data. Instead, it writes new versions of the data that it inserts or updates into new SSTables. There's never an in-place revision of data (which requires random I/O). Instead, the database simply writes the latest version of the data (updates or inserts) in new SSTables. This is a key factor in Cassandra's high performance.

Note   The database turns all updates to data into new SSTables that it writes sequentially to disk.

As the database updates data over time, it may end up with multiple versions of the same row, each in a different SSTable. Unique timestamps distinguish the versions.

Over time, Cassandra needs to access numerous SSTables to retrieve a complete row, as no single SSTable is likely to be storing the most up-to-date versions of all columns in that row. To improve performance, Casandra performs periodic compactions of the SSTables by merging the tables and getting rid of old, outdated versions of data. Compaction keeps the read speed from deteriorating as users update data over time.

During the compaction process, Cassandra does the following;

- Consolidates the SSTables

- Evicts all tombstones

- Merges keys

- Combines columns

- Creates a new index in the new SSTable (not a secondary index)

The compaction process creates a single complete row with the latest version of all columns in that row. The database writes the latest versions of each row to a new SSTable and deletes the old versions from the old SSTables once it satisfies all pending requests. Here are the practical consequences of the compaction process:

- An improvement in read performance due to the newly compacted SSTables

- A temporary rise in disk I/O and space usage since the database needs to temporarily maintain both the old and new SSTables

You can manually compact tables with the `nodetool compact` command. Better still is to set the compaction strategy for a table, either via a `CREATE TABLE` or an `ALTER TABLE` command.

Compaction Strategies

A compaction strategy determines how Cassandra selects SSTables for compaction and how it stores compacted rows in the new SSTables that it creates during compaction for holding up-to-date data. You can choose from the following three compaction strategies:

- SizeTieredCompactionStrategy (STCS)

- LeveledCompactionStrategy (LCS)

- TimeWindowCompactionStrategy (TWCS)

Note   There's also a DataTieredCompactionStrategy (DTCS), which is an alternative strategy for time series data, but it's deprecated in Cassandra 3.0 and later.

TWCS is the best strategy for compacting time series data. For non-time series data, you need to choose either STCS or LCS. The following sections explain the three compaction strategies you can choose from.

SizeTieredCompactionStrategy

Under STCS, the size of the SSTables acts as the trigger for a table compaction. STCS groups SSTables into unique buckets. The database compacts the SSTables once there are a predetermined number of same-sized tables. Cassandra compares each SSTable's size to the average of the SSTable size on a node.

By default, the compaction configuration property enabled for STCS is true, meaning that the database performs background compaction. The SSTable configuration property min_threshold sets the minimum number of SSTables that'll trigger a compaction. Once a table meets the min_threshold value, it's eligible for compaction. By default, the database performs a compaction operation when there are four similar-sized tables (160MB by default). Cassandra merges the same-sized tables into a single, large SSTable. As the database acquires several of the larger SSTables, it merges a set number of them into an even larger SSTable.

Note   STCS is the default compaction strategy.

The following is an example showing how to change a table's compaction property to SizeTieredCompactionStrategy:

```
cqlsh> ALTER TABLE users
       WITH compaction =
       'class' : 'SizeTieredCompactionStrategy', 'min_threshold' : 6 }
```

The min_threshhold table property enables you to set the minimum number of SSTables before the database performs a minor compaction. In this example, the min_threshold property specifies that at a minimum, the database should look for six SSTables of the same size before it triggers a compaction.

Note   A minor compaction involves specific tables in a keyspace.

Configuring STCS

In addition to the enabled and the min_threshold properties, you can configure the
following properties for STCS:

- bucket_high and bucket_low: This pair of properties provides the
  sizing formula that helps the database determine the size of a table
  for it to be considered for compaction. The database merges SSTables
  whose size falls within the following range:

  ```
  [average table size  X bucket_low] and [average table
  size X bucket_high]
  ```

  The default value of the bucket_high property is 1.5, and the
  bucket_low property has a default value of 0.5.

- log_all: Enables you to set up advanced logging for the cluster, and
  its default value is false.

- min_threshold and max_threshold: This set of properties determines
  the minimum (default value is 4) and the maximum (default value is 32)
  number of SSTables in a minor compaction.

- min_sstable_size: By default, the database groups SSTables into
  buckets by including in each bucket SSTables that vary in size by less
  than 50%. In cases where you have smallish SSTables, you can set
  the min_sstable_size property to define a size threshold (in bytes)
  for SSTables that are assigned to a bucket. The default value for this
  parameter is 50MB.

- *Tombstone-Related Compression Properties*: There are several
  tombstone-related compression properties.

  - only_purge_repaired_tombstones: The default value of this
    parameter is false, meaning that the database will allow purging
    tombstones from all SSTables, repaired as well as unrepaired. By
    setting this property to the value true, you require the database
    to purge tombstones only from repaired SSTables.

- `tombstone_threshold`: Sets the ratio of the tombstones that the database can garbage collect, compared to all contained columns. The default value for this parameter is 0.2. Once an SSTable exceeds this ratio (20%), it's eligible for tombstone compaction. Once the number of tombstones that the database can garbage collect exceeds 20%, Cassandra will compact just that table to purge the tombstones.

- `tombstone_compaction_interval`: The length of time after you create an SSTable before the database considers the table for tombstone compaction. The default value is 86,400 seconds (1 day).

- `unchecked_tombstone_compaction`: By default, this property is set to `false`, meaning that Cassandra will check for the eligibility of the table for tombstone compaction before running the compaction operation. By setting the property to `true`, you allow Cassandra to compact the tombstones without the prechecks.

When to Use (or Not to Use) STCS

STCS is ideal for write-intensive workloads. Since the merge process doesn't group data by rows, it's possible that versions of rows are spread over multiple tables, which slows down read speeds.

However, STCS doesn't evict stale and deleted data quickly since the compaction trigger is the size of the SSTables, which may grow too slowly over time, thus keeping old data in place. Over time, as SSTables get larger, since the database needs to store both the old and new SSTables during compaction, the database will require more storage.

Although LCS keeps related data in a small set of tables, if your data doesn't undergo too many modifications or inserts, STCS can also give you the same type of data sets without paying the write penalty that you incur with LCS.

STCS is more efficient than LCS when dealing with batched read and write operations. On the negative side, however, it requires a lot more disk space than LCS.

LeveledCompactionStrategy

LCS groups SSTables into levels where each level (L1, L2, L3 …) is 10 times larger than the preceding level. The database compacts SSTables in each level into progressively larger levels. To begin with, the database flushes the data in memtables to SSTables in the lowest level (L0). The database doesn't compact the tables in the L0 level; it merges these smallest sized SSTables with larger SSTables at the next higher level (L1).

The database compacts the SSTables in levels L2, L3, and so on into SSTables that are at least as large as the value you set for the LCS property `sstable_size_in_mb` (default is 160MB).

LCS improves read performance since Cassandra can determine the SSTables in each level it should check for the existence of row key data.

Configuring LCS

You can set the following properties for LCS, and all these properties have the same meaning as they do for STCS, as I described in the previous section.

- `enabled`
- `log_all`
- `tombstone_compaction_interval`
- `tombstone_threshold`
- `unchecked_tombstone_compaction`

In addition to these properties, you can also configure the sstable_size_in_mb property, which is unique to LCS. The sstable_size_in_mb property specifies the target size for tables when compacting with LCS. The database will try to keep the SSTable size (after compaction) the same or less than the value you specify for the sstable_size_in_mb parameter. However, if a partition is very large, the compaction will result in a larger SSTable than the value you specified with the sstable_size_in_mb parameter. As mentioned earlier, the default value for the sstable_size_in_mb property is 160MB.

Here's an example showing how to update a table to set the database compaction strategy to LeveledCompactionStrategy:

```
cqlsh> ALTER TABLE users WITH
         compaction = { 'class' :   'LeveledCompactionStrategy'};
```

When to Use (or Not to Use) LCS

LCS involves a tradeoff between read and write performance. LCS works well for read-intensive workloads (especially workloads that involve random reads) since it enables the database to retrieve the required data for queries from just one SSTable 90 percent of the time. The other times, it'll need to read just two SSTables.

LCS is ideal when there's a lot of data fragmentation. LCS evicts stale data more often than STCS, which means that deleted data eats up a lesser part of the SSTables. However, more frequent compaction operations mean there's more I/O pressure. Thus, LCS isn't generally a promising idea for write-intensive workloads due to the performance hit stemming from the higher amount of I/O operations. LCS's write penalty means that a high volume of writes could overwhelm the compaction operations.

LCS requires less space than STCS.

If it's critical for you to maintain a high read rate, you can use LCS. To overcome the resulting performance hit on the write side, you can add more nodes to your cluster.

TimeWindowCompactionStrategy

TWCS is the recommended compaction strategy for time series data and expiring TTL workloads. TWCS compacts SSTables according to a series of time windows. The database creates continuous time windows and during each active (most recent) time window, it compacts all uncompacted SSTables into larger SSTables using STCS compaction.

At the end of the time window, the database compacts (major compaction, since SSTables from more than one keyspace are involved) all the SSTables that fall in that time window into a single SSTable, using the SSTable maximum timestamp as a criterion. The database repeats the same compaction procedures with all the SSTables that it writes during the subsequent time windows.

Configuring TWCS

You can set the following compaction properties for configuring TWCS:

- compaction_window_unit: This property enables you to define the time unit for defining the bucket size. The default time unit is milliseconds, and you can set a value such as seconds or hours if you wish. Here's an example:

  ```
  compaction_window_unit = 'minutes',compaction_window_size = 120
  ```

- compaction_window_size: The number of units per time window (1, 2, 3 ...).

- log_all: Enables you to activate advanced logging for the entire cluster by setting the value true. The default value is false.

When to Use TWCS

TWCS is ideal for time series data stored in tables with a default TTL (time-to-live). This compaction strategy isn't good if you need to query time series data out of sequence.

Enabling and Disabling Compaction

By default, Cassandra enables background compaction. You can disable the background compaction by setting the enabled property to false through an ALTER TABLE statement, as shown here:

```
cqlsh> ALTER TABLE mytable
       WITH COMPACTION = {
       'class': 'SizeTieredCompactionStrategy',
       'enabled': 'false' }
```

The best practice regarding compaction is to use the default setting of enabled (true) and let the database compact data.

Configuring Global Compaction Properties

You can configure the following three global compaction parameters:

- snapshot_before_compaction

- concurrent_compactors

- compaction_throughput_mb_per_sec

I explain these properties in the following sections.

The snapshot_before_compaction Property

The snapshot_before_compaction property determines whether the database takes a snapshot of data before performing a compaction operation. As with all of Cassandra's snapshots, you're responsible for deleting older snapshots so they don't consume too much space.

By default, the snapshot_before_compaction property is set to false.

The concurrent_compactors Property

During a long-running compaction, a large number of small SSTables can accumulate, which will adversely impact read performance. To avoid this deterioration in the read performance, you can configure the concurrent_compactors parameter to set the number of concurrent compaction processes that can run at any time. This number doesn't include any validation compactions for anti-entropy repair.

By default, the concurrent_compactors property is set to the lower of the number of disks or the number of cores, with a minimum of two and maximum of eight per CPU core.

Since performing simultaneous compactions increases the use of disk space, make sure that you have sufficient free disk space before raising the value of the concurrent_compactors parameter.

Note   Before you adjust the concurrent_compactors property, throttle the speed of the compaction by configuring the compaction_throughput_mb_per_ sec property.

The sstable_preemptive_open_interval_in_mb Property

The database can preemptively open the SSTables it is compacting before it completes its writes to the SSTables. The goal is to smoothly transfer reads between the before and after compaction versions of the SSTables by minimizing the movement of data into and out of the cache, as well as keep the hot rows in place. The default value is 50MB.

Throttling the Compaction Speed

Sometimes you may realize that a compaction is running too fast (or too slow). You can configure the `compaction_throughput_mb_per_sec` property to throttle compaction to a specific rate (measured in MB) per second. This compaction property controls the speed with the database inserts data. The faster the rate of data insertion, the higher the speed with which the database must perform the compaction so it can keep the SSTable count low.

The default value for the `compaction_throughput__mb_per_sec` parameter is 16, meaning that the rate of compaction is 16 times the write throughput, which is measured in MB/second.

You can disable compaction throttling by setting the `compaction_throughput__mb_per_sec` property to 0. The recommended value is 16-32. The compaction speed is divided among all compactors. If you have eight concurrent compactors and 16 as the value for compaction throughput, you only get 2MB/s per compactor. On spinning disks it is advisable to have a really low number of concurrent compactors (ex: 2); on SSD, you can raise the number of compactors to a level your disks can keep up with.

Note   You can have the database issue a warning when compacting a partition that's larger than the value you set by configuring the `compaction_large_partition_warning_threshold_mb` property. The default value for this parameter is 100.

Setting the Compaction Strategy

You set the compaction strategy at the table level. You can set the compaction properties either when you create a table or by altering the table. You specify the compaction `class` with the `compaction` option inside a `CREATE TABLE` or an `ALTER TABLE` statement, as shown here:

```
compaction = {
    'class' : 'compaction_strategy_name'
    [, 'subproperty_name' : 'value',...]
}
```

Although you can disable compaction, it's never a good idea to do so because it may lead to the propagation of zombies. A zombie, if you recall from Chapter 5, is a deleted but persistent record. Sometimes a tombstoned record will have been deleted from other nodes of the cluster except for an unresponsive node, which doesn't immediately receive the tombstones since it's down. If the database deletes the tombstoned records from the rest of the cluster before the unresponsive node comes back up, Casandra treats the (supposed to be deleted) record on this node as new data and sends it to the other nodes.

As mentioned, a grace period for the unresponsive nodes to recover helps prevent the reappearance of the zombie records. The table property `gc_grace_seconds` sets the grace period for a tombstone, and its default value is ten days (864,000 seconds). Once the tombstone's grace period ends, the database deletes the tombstones during the compaction process.

Getting and Setting the Compaction Thresholds

Run the `nodetool getcompactionthreshold` command to find the minimum and maximum compaction thresholds for a table. You must specify the keyspace and the table name when you run this command.

```
$ sudo nodetool getcompactionthreshold cycling cyclist_name
Current compaction thresholds for cycling/cyclist_name:
 min = 4,  max = 32
$
```

Note   A major compaction involves all tables in a keyspace. A minor compaction involves only some of the tables in a keyspace.

You can set the compaction thresholds with the nodetool's `setcompactionthreshold` command, which enables you to set the min/max compaction thresholds for a table. The term "threshold" refers to the number of similar-sized SSTables that should be in the database before the database schedules a minor compaction.

The `setcompactionthreshold` command has the following options:

- `keyspace`: Keyspace name

- `table`: Table name

- `minthreshold`: Minimum number of SSTables to trigger a minor compaction (when using STCS or DTCS).

- `maxthreshold`: The maximum number of SSTables in a minor compaction (when using STCS or DTCS).

Here's an example showing how to set the compaction threshold for a table:

```
$ nodetool setcompactionthreshold cycling cyclists 4 16
```

In this example, *cycling* refers to the keyspace and `cyclists`, the table.

As explained earlier, both STCS and DTCS allow you to configure a minimum and maximum number of SSTables that should be compacted during a minor compaction through the configuration if the min_threshold and max_threshold properties.

Logging the Compaction Activity

You can set up extended logging of the compaction activity in a database by setting the `log_all` subproperty to `true`. The `log_all` subproperty is under the `compaction` property when you create or alter a table. The `log_all` property for a table activates advanced logging for the entire cluster. While you do this at the table level, once you configure extended logging for a single table, the database automatically collects detailed information for compacting all the tables on all nodes in the cluster.

Once you set up extended logging of compaction activity, the database creates an additional compaction-related file named `compaction-%d` under the $CASSANDRA_HOME/ `logs` directory, with %d acting as a sequential number.

Following the setting up of extended compaction logging, the database collects information about the following compaction events:

Enable: Lists all the SSTables that the database has flushed.

```
{"type":"enable","keyspace":"es":
...
}
```

Flush: Logs flush events from memtables to SSTables.

```
{"type":"flush","keyspace":"test","table":"t","time":1470083335639,"tables":
...
}
```

Type: compaction: Logs compaction events

```
{"type":"compaction","keyspace":"test","table":"t","time":1470083660267,
"start":"1
...
}
```

Pending: Lists the number of pending tasks for compaction.

```
{"type":"pending","keyspace":"test","table":"t","time":1470083447967,
"strategyId":"1","pending":100}
```

Testing the Efficacy of a Compaction Strategy

When you're planning to change your compaction strategy, you can first test the impact of the new strategy on a node that's not officially part of the cluster. You start up a new node in the *write survey mode*, which adds the node to the cluster but doesn't formally make it a part of the ring.

You can, if you wish, take one of the nodes out of the cluster and bring it back in the write survey mode to test the impact of the changes on the read performance. Once you start the node in the write survey mode, you can benchmark the read operations on that node.

To start a node in the write survey mode, set the `write_survey` option to `true` by using either of the following two methods:

```
JVM_OPTS="$JVM_OPTS -Dcassandra.write_survey=true         /* for package
installations, add this line to the cassandra-env.sh file.
$ cassandra -Dcassandra.write_survey=true                 /* for tarball
installations, start the new node in this mode.
```

You can also start a node in the write survey mode by setting the property in the `$CASSANDRA_HOME/conf/jvm.options` file, as shown here (by default, the -Dcassandra.write_survey property is commented. You need to remove the comment). Here's the relevant portion of the `jvm.options` file:

```
# For testing new compaction and compression strategies. It allows you to
experiment with different
# strategies and benchmark write performance differences without affecting
the production workload.
-Dcassandra.write_survey=true
```

Enabling and Disabling Autocompaction

You can enable and disable auto compaction for one or more tables with the following commands:

```
$ nodetool enableautocompaction cycling cyclist_name;
$ nodetool disableautocompaction cycling cyclist_name;
```

Viewing the History of All Compactions

You can view the history of all compaction activity in the database with the `nodetool compactionhistory` command:

```
$ nodetool compactionhistory
Compaction History:
id                                  keyspace_name columnfamily_
name    compacted_at              bytes_in bytes_out rows_merged
f6227aa0-81e8-11e7-a4ad-89801d899afb cycling       cycle
2017-08-15T11:38:41.994 111       55        {1:1, 2:1}
```

```
ef3d57a0-81e8-11e7-a4ad-89801d899afb cycling        cyclist_name
2017-08-15T11:38:30.426 243        243        {1:5}
...
```

There are seven columns in the output of this command. The bytes_in and the bytes_out columns help you figure out the efficacy of compaction by comparing the size of the SSTables before and after compaction.

The notations in the output such as {1:5} refer to how Cassandra chose the rows for compaction and stands for {tables:rows}. For example, {1,5} means that the database took one row from five SSTables to create a new compacted SSTable. Similarly, {1:3,3:1} means that the database took three rows from an SSTable and combined with a single row from three tables to create a compacted SSTable.

The nodetool compactionstats command shows the status of all pending and ongoing compaction operations. You can check the percentage of compaction operations completed thus far by looking under the PROGRESS column.

Compressing Data to Save Storage Space

You can save storage space by having Cassandra compress SSTables on disk. In addition to providing gains on the storage front, compression offers a slight improvement in performance. DataStax estimates that, depending on the type of data, compressing could lead to a 25-33% drop in storage, a 25-35% improvement in read performance, and a slight (5-10%) increase in write performance.

When Cassandra needs to read compressed data, it locates the rows in the SSTable index and decompresses the necessary row chunks.

The best candidates for compression are tables with similar sets of columns, where the rows contain the same number of columns. The more similar the data is in various rows, the more space you can save from compressing the data, since the compression factor will be higher.

Configuring Compression

Cassandra enables compression by default, and you can configure various compression options, including the compression algorithm the database must use during compression. You can do this either when you create the table, or later, by altering the table. The following syntax statement shows the four compression properties you can configure:

```
compression = {
   ['class' : 'compression_algorithm_name',
     'chunk_length_kb' : 'value',
     'crc_check_chance' : 'value',]
   | 'sstable_compression' : '']
}
```

The most important compression property is the compression class, which specifies the compression algorithm the database must be use during compression. You can choose from the following compression classes:

- LZ4Compressor

- SnappyCompressor

- DeflateCompressor

Note   In addition to the LZ4, Snappy, and Deflate compression classes, you can also implement your own compression class with the org.apache.cassandra.io.compress.ICompressor interface.

When you choose a compression algorithm, you're choosing between the conflicting goals of conserving storage space and the database's read performance. Commonly, the savings in storage space you gain by compresssing data is inversely related to the speed with which the database can uncompress the data while reading it.

Both Snappy and Deflate offer a higher compression ratio, meaning you save more in storage space when using these algorithms, compared to using the LZ4 algorithm. However, LZ4 offers the fastest decompression speed, which means a higher read performance when querying compressed data.

The default compression algorithm is the LZ4Compressor. To specify a compression algorithm for compressing a table, you specify the *compression algorithm class* for that compression algorithm, as shown here:

```
CREATE TABLE customers (
...
    PRIMARY KEY (id)
      )
    WITH compression = { 'class' : 'SnappyCompressor' };
```

In addition to specifying the compression algorithm, you can also configure the following compression-related properties when creating or altering a table:

- chunk_length_kb: By default, the value of this property is set to 64KB. This is the size of the compressed blocks when Cassandra compresses an SSTable. By specifying a higher value for this parameter, you can increase the compression ratio, but the database must then read more data from a disk for every read operation. The default value is fine for most use cases, but if the database is reading a large amount of data from disk at once, you can raise the value of this parameter.

- crc_check_chance: This property determines the probability that the database will check the checksums of compressed data blocks during a read operation. Checksums help detect corruption and prevent the database from propagating corrupted replicas of data to other nodes. The default value of the crc_check_chance property is 1, meaning the database will always check the checksums of data during reads (a probability of 1 means that an event will always happen). You can disable checksum checking entirely by setting this parameter to 0. The lower the value of the parameter, the less likely it is that Cassandra will perform a checksum of the data during reads.

When Cassandra Compresses Data

When you configure compression for an existing table with the ALTER TABLE statement, Cassandra will start compressing all new SSTables for that table. However, Cassandra doesn't immediately compress all existing SSTables. Cassandra will compress the existing SSTables when it performs the next compaction of data in the database.

You can modify this default behavior of Cassandra for compacting data by forcing it to rewrite the existing SSTables, compressing the tables in the process. You can do this by executing either the nodetool upgradesstables or the nodetool scrub commands.

The nodetool upgradetsstables command has the following syntax:

```
$ nodetool <options> upgradesstables
 ( -a | --include-all-sstables )
 -- <keyspace>  <table> ...
```

The nodetool upgradetsstables command rewrites SSTables and during that process also compresses them. You can run this command if you want the database to immediately start compressing data following your configuring of compression for an SSTable. You can also run this command when you wish to modify the compression algorithm for a table.

You can make Cassandra compress a table for which you've configured compression by running the nodetool scrub command. This command will also rebuild the SSTables in a node, just as the nodetool upgradesstables command, but it needs to create a snapshot before the table rebuild. The nodetool upgradesstables command is a better choice than nodetool scrub since you don't need to create a snapshot before the database compresses the SSTables.

Modifying the Compression Algorithm

The default compression algorithm is LZ4Compressor. You can change the compression algorithm when creating a table, as shown here, by configuring the compression parameter class to a different compression class (DeflateCompressor in this example):

```
CREATE TABLE MyTable (
    ...
    PRIMARY KEY (user_id)
    )
    WITH compression = { 'class' : DeflateCompressor };
```

You can modify any compression property after creating a table, including the `compression class` property. The following example shows how to modify the compression algorithm `class` attribute (to `DeflateCompressor`):

```
ALTER TABLE customers
WITH compression = { 'class' : 'DeflateCompressor', 'chunk_length_in_kb' : 6
```

Testing the Efficacy of Compression

During your initial forays into compression, you may want to figure out the impact of alternative compression strategies without impacting the production database performance. Cassandra allows you to test your compression strategies by letting you start the database in the write survey mode.

The "Testing the Efficacy of a Compaction Strategy" section in this chapter describes how to use Cassandra's write survey mode to test out your compression plans.

Starting a new node in the write survey mode helps test new compression strategies and benchmark the write performance. To check the impact of compression strategies on read performance, you can take out a node from the cluster and benchmark (using the `cassandra-stress` tool, which I explain in this chapter) the performance of its read operations, using that node as a standalone server.

Turning Off Compression

By default, Cassandra compresses data. You can disable compression when creating a table by specifying the compression parameter `enabled` to the value `false`, as shown here:

```
CREATE TABLE MyTable (
    ...
    PRIMARY KEY (user_id)
    )
    WITH compression = { 'enabled' : false };
```

You can disable compression for an existing table by specifying the sstable_compression option and giving it the value (' '), as shown here:

```
cqlsh> ALTER TABLE cycling.cyclist_name
       WITH COMPRESSION = {'sstable_compression': ' ');
```

Improving Write Performance with Memtable Thresholds

As you're aware, Cassandra flushes memtables to disk (to SSTables) when it exceeds either the commit log threshold or the memtable cleanup threshold. There's no fixed rule regarding setting the commit log and memtable thresholds.

You can raise the memtable threshold if your write loads include high updates on a small set of data or if there's a continuous steam of writes. If you keep the memtable threshold low when there are numerous writes, it will result in inefficient compaction.

Cassandra uses the setting of the memtable_cleanup_threshold parameter (in cassandra.yaml) to determine when it should flush memtables. The default value for this parameter is 1/(memtable_flush_writers + 1). However, this parameter only signifies the ratio Cassandra uses for computing the space amount for an automatic memtable flush. Cassandra determines when to flush a memtable to disk in the following way:

- The memtable_heap_space_in_mb parameter determines the amount of on-heap memory allocated for memtables. The default value for this parameter is 0.25 times the heap size. The memtable_offheap_ space_in_mb parameter determines the amount of off-heap memory allocated for memtables and its default value is also 0.25 times the heap size. Add the value of the memtable_heap_space_in_mb parameter to the value of memtable_offheap_space_in_mb parameter.

- Multiply the total from the preceding line by the value of the memtable_cleanup_threshold parameter. This will give you the threshold value in MB.

- When the combined memory used by all memtables on a node exceeds the threshold value, the database flushes the largest memtable on this node to disk.

The memtable_flush_writers parameter denotes the number of memtable flush writer threads. The default value for this parameter, which the database uses for computing the value of the memtable_cleanup_threshold parameter, is determined in the following manner:

The number of disks or the number of cores on a node, with the minimum value being 2 and the maximum being 8.

The following is an example that shows how the database determines when to flush a memtable to disk.

1. Assume that you have set the node's memtable_flush_writers parameter to 8.

2. The default value of the memtable_cleanup_threshold parameter is 1/memtable_flush_writers +1, so in this case, it works out to 1/(8+1) = 0.11.

3. Assume that you've set both the memtable_heap_space_in_mb and the memtable_offheap_space_in_mb parameters to 2000.

4. The memtable_cleanup_threshold parameter's value is 0.11, from Step 2. The sum of the memtable_heap_space_in_mb and the memtable_offheap_space_in_mb parameters is 2000+2000=4000. The space threshold for memtable cleanup then is 4000 - 4000*0.11=3636MB.

5. When the combined amount of space used by the memtables on this node exceeds 3636MB, Cassandra flushes the largest memtable to disk.

The larger the value of the memtable_cleanup_threshold parameter, the less frequently the database will flush the memtables to disk. This also means fewer SSTables, which in turn means a smaller level of compaction activity.

Tuning the JVM

Everything runs inside a JVM in Cassandra. Cassandra uses the JVM heap to optimize read performance. To do this, it keeps components such as Bloom filters, the partition key cache, the SSTable index summary, and the partition summary inside the heap.

In addition to the Java heap memory, Cassandra uses memory for supporting the page cache, the Bloom filters, and cached rows.

It's important to learn how to tune the JVM. The key things to know in this regard are

- How Java garbage collection works

- The ideal garbage collector to use

- Allocating the Java heap size

Configuring garbage collection and allocation of the correct amount of memory to the JVM (heap) are critical to high performance. Let's learn about both in the following sections.

Java Garbage Collection

Garbage collection is how the JVM removes older objects from memory. You have a choice of different garbage collectors. There are several types of regions inside a Java memory heap, such as young and old regions, and within each memory region there are smaller subregions as well. When the JVM needs to clear up a memory region that's full, it pauses the running operations. Your goal when tuning Java garbage collection is to keep the pauses to a minimum. You should not only seek to minimize the garbage collection-induced pauses to a minimum, but also to reduce the number of pauses.

Selecting the Right Garbage Collector

The two available Java garbage collectors are the Concurrent Mark Sweep (CMS) and the G1 (Garbage First) garbage collector. Here's how you decide between the two garbage collectors:

- If you're working with heap sizes smaller than 14GB and you're dealing with the same workloads all the time, CMS is better.

Note   With the next release of Java (Java 9), CMS will be officially obsolete.

- If you're dealing with heap sizes larger than 14GB and the database is processing diverse types of workloads, use the G1 garbage collector.

You configure the Java garbage collector by editing settings in the $CASSANDRA_HOME/ conf/jvm.options file. The jvm.options file, which I first explained in Chapter 2, helps set several properties such as the startup parameters, general JVM settings, heap settings, and garbage collection-related settings.

The default garbage collector is CMS, and you can view the settings for this collector in the java.options file under the GC SETTINGS section, as shown here:

```
#   GC SETTINGS  #
################
### CMS Settings
-XX:+UseParNewGC
-XX:+UseConcMarkSweepGC
-XX:+CMSParallelRemarkEnabled
-XX:SurvivorRatio=8
-XX:MaxTenuringThreshold=1
-XX:CMSInitiatingOccupancyFraction=75
-XX:+UseCMSInitiatingOccupancyOnly
-XX:CMSWaitDuration=10000
-XX:+CMSParallelInitialMarkEnabled
-XX:+CMSEdenChunksRecordAlways
# some JVMs will fill up their heap when accessed via JMX, see
CASSANDRA-6541
-XX:+CMSClassUnloadingEnabled
```

If you want to switch the garbage collector from the default CMS collector to the newer G1 collection, first you must comment everything under the CMS Settings section. Following this, edit the G1 settings and uncomment the -XX:+UseG1GC property, as shown here:

```
### G1 Settings (experimental, comment previous section and uncomment
section below to enable)
## Use the Hotspot garbage-first collector.
-XX:+UseG1GC                        => By default, this property is
commented
#
```

Here are the G1 settings you can configure:

- XX:MaxGCPauseMillis: This is the key GI garbage collector tunable property. The lower the pause target, the lower the throughput. The lowest setting for the MaxGCPauseMillis property is 200 ms, which also happens to be the default value for this property. By raising the value from 200 ms to 1000 ms, you can increase the throughput. You must try to keep this parameter's value smaller than the timeouts in the cassandra.yaml file.

  ```
  ## Have the JVM do less remembered set work during STW, instead
  ## preferring concurrent GC. Reduces p99.9 latency.
  -XX:G1RSetUpdatingPauseTimePercent=5
  #
  ```

- -XX:InitiatingHeapOccupancyPercent: Setting this parameter saves the CPU time in Java heaps that are large (greater than 16GB). The JVM does this by holding off the scanning of the memory regions to see if they're full, until the heap gets to a set percentage full. If, for example, you set the heap occupancy percent to 70%, it means that when the heap gets 70% full, the JVM starts scanning the memory regions. The default value for this parameter is 40%.

I don't show any configuration properties for the CMS collector since it's on its way out.

Setting the Heap Size

You can configure the Java heap size by setting the properties listed under the HEAP SETTINGS section in the jvm.options file. When you are using G1 garbage collector, you need to configure the MAX_HEAP_SIZE property with the -Xmx property. You set the minimum heap size with the -Xms property and the maximum heap size with the -Xmx property.

By default, the -Xms and -Xmx properties are commented, and Cassandra automatically figures the value of the MAX_HEAP_SIZE property with the following simple heuristic:

max(min(1/2 ram, 1024MB), min(1/4 ram, 8GB))

That is, it estimates half the RAM and caps it at 1024MB. It also calculates the quarter value of the RAM and caps it at 8MB. It then picks the higher (max) of the two values.

In a production database, it's a best practice to set custom values for the -Xmx and the -Xms properties. When setting the maximum value of the heap, the recommended value for the MAX_HEAP_SIZE property is to set it as high as possible, up to a value of 64GB.

Another (probably essential) best practice is to set the min and max heap sizes to the same value, as shown here:

```
-Xms16G
-Xmx16G
```

Setting the minimum and maximum heap sizes to the same value helps the database avoid long garbage collection pauses (also called stop-the-world pauses) when the JVM clears up space in the heap regions by removing stale objects. Instead, the JVM locks the memory you specify for the heap when the server starts up to keep parts of it from being swapped to disk.

Configuring Garbage Collection Logging

It's a best practice to enable GC logging and by default it is enabled. You can check the GC logs to see the size of the heap usage by a node, which helps you adjust it as you go along. Here's the relevant portion from the `java.options` file:

```
### GC logging options -- uncomment to enable
-XX:+PrintGCDetails
-XX:+PrintGCDateStamps
-XX:+PrintHeapAtGC
-XX:+PrintTenuringDistribution
-XX:+PrintGCApplicationStoppedTime
-XX:+PrintPromotionFailure
#-XX:PrintFLSStatistics=1
#-Xloggc:/var/log/cassandra/gc.log
-XX:+UseGCLogFileRotation
-XX:NumberOfGCLogFiles=10
-XX:GCLogFileSize=10M
```

As you can tell, the properties shown here help you configure various aspects of garbage collection logging, such as the location of the log, log file rotation, the size of the log files, and the number of log files to store.

Using the nodetool proxyhistograms and the tablehistograms Commands

There are two good nodetool commands that help you identify performance issues in a cluster: `nodetool proxyhistograms` and `nodetool tablehistograms`. Both commands present performance statistics captured by the database in the form of histograms, hence the command names.

The `nodetool proxyhistograms` command is helpful in identifying performance issues. The output of this command shows read, write, and range request latencies in a cluster, for which a node acted as the coordinator. This command helps you identify general performance issues in a cluster.

The `nodetool tablehistograms` command lets you focus on the performance of a specific table. You can view the write and read latency and the partition size of tables with this command.

Summary

This chapter presented several key Cassandra performance tuning strategies such as compaction and compression. Performance tuning often involves tradeoffs; you can conserve storage through compression, but you must pay for it with higher I/O requirements. Similarly, increasing the Java heap size will decrease query latencies, but reduces the amount of memory available for other uses in the system.

The `cassandra-stress` tool is simple to use and is highly useful in testing the impact of potential configuration changes such as a compaction strategies, compression, etc. Stress testing helps you evaluate the ability of a cluster to handle diverse types of workloads and helps you figure out if you need to add more nodes to your cluster.

CHAPTER 12

Securing Cassandra

Securing a Cassandra cluster involves a varied set of tasks including authentication, authorization, and encryption. This chapter describes how you manage the following security-related tasks:

- *Authentication*: How you allow applications and users to log into the cluster.

- *Authorization*: Deals with the granting of permissions to access a database or database objects such as tables and materialized views.

- *Encryption*: Refers to the use of the Secure Socket Layer (SSL) to secure communications between clients and Cassandra databases, and among a cluster's nodes.

- *Firewalls*: Managing firewall port access involves knowing which ports you must keep open.

Configuring Authentication

In most databases, it's typical to use the terms *users* and *roles* to refer to different entities. Users are login accounts and roles encapsulate sets of privileges on various objects that you assign to users.

Cassandra bases all authentication and authorization on roles. You execute the CREATE ROLE command to create database roles.

© Sam R. Alapati 2018
S. R. Alapati, *Expert Apache Cassandra Administration*, https://doi.org/10.1007/978-1-4842-3126-5_12

> **Note**   All authorization and authentication is through database roles. Although Cassandra continues to offer the `CREATE USER,` `ALTER USER,` `DROP USER,` and `LIST USERS` commands, you mustn't use them because they're deprecated in the current release. Use the `CREATE ROLE,` `ALTER ROLE,` `DROP ROLE,` `LIST ROLES,` and `LIST_PERMISSIONS` commands instead.

Cassandra still offers the `CREATE USER` command to create new database user accounts, but this command is deprecated and is there just for backwards compatibility purposes.

Cassandra doesn't formally use the term *users* to refer to login accounts any longer. Therefore, a (login) role is a synonym for a user.

By default, Cassandra doesn't require authentication for someone to log into a cluster. That is, you can simply type in *cqlsh* without any credentials and gain access to the cluster.

Creating Roles

Cassandra comes with a built-in role named *cassandra* and the password is *cassandra* as well. This default role *cassandra* has administrator privileges. You can also create additional login roles and grant them the administrator privilege. The best practice is to create alternative administrator roles, change the default password of the default role *cassandra*, and never use this role for anything again.

You can create roles only by logging in as a role (could be the default role *cassandra* or any other role you've created) with administrative privileges.

When you first log in as the default role *cassandra* and attempt to create a role, you may receive the following error;

```
$ cqlsh 192.168.159.129 -u cassandra -p cassandra
Connected to Test Cluster at 192.168.159.129:9042.
[cqlsh 5.0.1 | Cassandra 3.10 | CQL spec 3.4.4 | Native protocol v4]
Use HELP for help.
cassandra@cqlsh> create role newrole with password 'newrole123';
SyntaxException: line 1:20 mismatched input 'withpassword' expecting EOF
(create user newuser [withpassword]...)
```

```
cassandra@cqlsh> create role newrole with password 'newrole123';
InvalidRequest: Error from server: code=2200 [Invalid query] message="org.
apache.cassandra.auth.CassandraRoleManager doesn't support PASSWORD"
cassandra@cqlsh> list users;
Unauthorized: Error from server: code=2100 [Unauthorized] message="You have
to be logged in and not anonymous to perform this request"
cassandra@cqlsh>
```

The reason you see this error is that you haven't yet configured authentication in this database. By default, the authentication option in the `cassandra.yaml` file is set to

```
authenticator: AllowAllAuthenticator
```

`AllowAuthenticator` is the default authentication backend. If you set `AllowAuthenticator` as the value for the authenticator property, you disable authentication in the database. Cassandra won't perform any checks and will allow anyone to log in without a challenge.

The alternative to `AllowAuthenticator` as the authentication backend is `PasswordAuthenticator`. This backend will authenticate users with credentials that the database stores in the `system_auth.credentials` table.

Therefore, even when you log into the database as the default role *cassandra* (which has the administrator rights), the database tells you that "you have to be logged in and not (be) anonymous to perform this request" when you try to create a role. The `authenticator: AllowAllAuthenticator` option allows all users and doesn't check if you have logged in. The solution here is to configure authentication for the database, as I explain in the following section.

Configuring Authentication

To enable you to log in as an administrator and create roles or perform other role-related tasks, configure authentication by executing the following steps.

1. Change the authentication option in the `cassandra.yaml` file to `PasswordAuthenticator`, as shown here:

    ```
    authenticator: PasswordAuthenticator
    ```

    The change you make here forces Cassandra to require a role name and password when a client connects to the cluster.

2. Restart the database.

3. Login to cqlsh using the credentials for the default superuser *cassandra*.

    ```
    $ cqlsh -u cassandra -p cassandra
    ```

4. The system keyspace *system_auth* stores the role credentials. If you've only the default single replica for this keyspace, you can lose access to the cluster if that replica becomes unavailable. Therefore, to enhance availability, increase the replication factor for the *system_auth* tablespace to a higher value such as 3 or 5 per datacenter, as shown here:

    ```
    cqlsh> ALTER KEYSPACE "system_auth"
             WITH REPLICATION = {'class' :
             'NetworkTopologyStrategy', 'dc1' : 5,
             'dc2' : 3};
    ```

5. To ensure that the replication factor changes are enforced throughout the cluster, run the `nodetool repair` command for the *system_auth* keyspace.

    ```
    $ nodetool repair system_auth
    ```

6. Restart the database.

    At this point, you've configured authentication for the database and you can now login as `cassandra/cassandra` to perform role- and privilege-related administrative tasks.

    Once you've configured authentication as shown here, you must always specify the credentials for a superuser. If you try to log in as before without the credentials, you'll receive an error.

    ```
    $ cqlsh 192.168.159.129
    Connection error: ('Unable to connect to any servers',
    {'192.168.159.129': AuthenticationFailed('Remote end
    requires authentication.',)})
    $
    ```

Speeding Up the Credentials Authentication Process

You can speed up the credentials authentication process in a busy database by configuring the following two options in the `cassandra.yaml` file:

- `credentials_validity_in_ms`

- `credentials_update_interval_in_ms`

The default value for both parameters is 2000 ms. You can raise the value of one or both parameters to decrease the overhead of frequent requests for authenticating credentials.

Creating Roles

Once you configure authentication, you're ready to create roles with the superuser account. Here you use the default superuser account *cassandra* to create the users.

You run the `CREATE ROLE` command to create a role. Here's the general syntax of the `CREATE ROLE` command:

```
(CREATE | ALTER | DROP ) role_name
 [WITH (LOGIN = true | SUPERUSER = true | password = 'password')];
```

The following is a brief explanation of the key role properties. By default, the database sets the `SUPERUSER` and `LOGIN` properties to `false`, and the `password` property defaults to `null`.

- SUPERUSER: Can execute all CQL commands. If you set this property to `true`, the rule is granted the `AUTHORIZE`, `CREATE`, and `DROP` privileges on all roles.

- LOGIN: If set, the database allows this role to log in with a password to run CQL statements. You set this property to create login accounts for the `passwordAuthenticator` backend or for internal authentication.

- PASSWORD: Represents the password that you set. Cassandra's internal authentication requires a password. You must enclose the password in single quotes.

You can create roles to define a set of permissions that you can then assign to other roles and map to external users. Roles also enable you to create login accounts for internal authentication. The best practice when using internal authentication is to create separate roles for assigning permissions and for the login accounts.

Here are the steps to create a role.

1. Log in to cqlsh as the superuser *cassandra*.

    ```
    $ cqlsh 192.168.159.129 -u cassandra -p cassandra
    Connected to Test Cluster at 192.168.159.129:9042.
    [cqlsh 5.0.1 | Cassandra 3.10 | CQL spec 3.4.4 | Native
    protocol v4]
    Use HELP for help.
    ```

2. Execute the CREATE ROLE <rolename> WITH PASSWORD <password> command, as shown here:

    ```
    cassandra@cqlsh> create user 'test' with password
    'test123';
    ```

    The list roles command shows all roles in the database. It also shows which of the roles have the super user privilege. Here's an example:

    ```
    cassandra@cqlsh> list roles;
     name      | super
    -----------+-------
     cassandra |  True
          test | False

    (2 rows)
    cassandra@cqlsh>
    ```

In this case, the default role *cassandra* has the superuser column value set to True, and the new role *test* has its the superuser column set to False. Therefore, everything is as it should be.

By default, the LOGIN property in the CREATE ROLE statement has the value False. When you're creating a login role, you must set this property to True.

You can view the roles in a database by querying the `system_auth.roles` table, shown here:

```
cassandra@cqlsh> select * from system_auth.roles;
 role           | can_login | is_superuser | member_of          |
salted_hash
---------------+-----------+--------------+--------------------+
    cassadnra1 |      True |         True |               null |
$2a$10$.b52eNZqZdouserevzYgpuecaxpdmc/QRdIoUIeG73a6UDEsBIdae
       manager |      True |        False | {'cycling_admin'}  |
$2a$10$vkujDzBblNqLUSGLdhW26OIW/ias9Vh6JA3sU4pq9uXES30cK735.
          test |      True |        False |               null |
$2a$10$ukMpmnzdHn8xy/7krucrdeYtblF8XHmVTvv1qKldVrGLxvLTWBxaC
         test1 |      True |        False |               null |
$2a$10$3CQTIi09zOOT5v2SaCxH9ekE8ZTaAGZSG6owkziZIT3ylzMDMCrhW
     cassandra |      True |         True |               null |
$2a$10$nRxFfMOAqxVQYkFQmEMIZ.B1.9.opcND/8.LtwyqZLgJY91eiS3Zm
 cycling_admin |     False |        False |               null |
  null

(6 rows)
cassandra@cqlsh>
```

The `system_auth.roles` table stores role-related information such as the role names, whether the role is a superuser, and whether you can use the role for logging into the database. In addition to this table, Cassandra stores role-related information in the following tables:

- `system_auth.role_members`: Stores roles and role members.

- `system_auth.role_permissions`: Stores the role, the resource, and the permissions that the role has to access the resource

- `system_auth.resource_role_permissons_index`: Stores the roles and the permissions granted to the roles

Changing a Password

You can modify the password for a role with the `ALTER ROLE` command:

```
cassandra@cqlsh> alter role 'newsuper' with password 'newsuper2';
```

Dropping a Role

You can drop a role with the DROP ROLE command:

```
cassandra@cqlsh> drop role test;
```

Handling the Superuser Account

Besides the default superuser account *cassandra*, you can create additional superuser accounts by specifying the superuser option when you create the user, as shown here:

```
cassandra@cqlsh> create role 'newsuper' with password 'newsuper1'
            ... superuser;
cassandra@cqlsh> list roles;
```

```
 name       | super
------------+-------
 cassandra  |  True
  newsuper  |  True
      test  | False
(3 rows)
cassandra@cqlsh>
```

The best practice concerning handling of the default superuser account *cassandra* is to create a custom administrator account and not use the default *cassandra* account after that. Create a user with with the superuser option. Since a superuser can only manage roles by default, grant access to all keyspaces to that user with the following command:

```
cqlsh> grant all permissions on all keyspaces to newsuper;
```

Configuring Authorization: Granting Privileges on Resources

Configuring authentication will limit access to the cluster. Configuring authorization is how you control access to various database objects such as keyspaces and tables. You can execute the GRANT command to grant a role privileges on a database resource such as a keyspace, table, or a function.

Let's learn how to grant privileges to login roles, starting with the granting of specific privileges on objects such as keyspaces and tables. Later, I show how to grant the privileges to roles, which enables you to implement a role-backed access control (RBAC) system.

Note   By default, Cassandra doesn't enforce any restrictions on a user's ability to perform operations in the database.

In the following example, you first try to grant the SELECT permission on a keyspace to a role:

```
cassandra@cqlsh> grant select permission on keyspace "cycling" to 'test';
ServerError: java.lang.UnsupportedOperationException: GRANT operation is
not supported by AllowAllAuthorizer
cassandra@cqlsh>
```

This GRANT command fails because the default authorizer, which happens to be AllowAllAuthorizer, doesn't support the GRANT operation. In the cassandra.yaml file, you see the following:

```
authorizer: AllowAllAuthorizer
```

AllowAuthorizer is Cassandra's default authorization backend and it controls access and provides permissions. Using this default authorization backend will disable authorization, meaning Cassandra allows any user to perform any action in the database.

CassandraAuthorizer is the alternative authorization backend. This one stores all permissions in the system_auth.permissions table. To configure authorization, edit the cassandra.yaml file and specify CassandraAuthorizer as the authorizer, as shown here:

```
-authorizer: org.apache.cassandra.auth.CassandraAuthorizer
```

The setting of the authorizer as shown here enforces authorization in the database. The database will now restrict access based on which role logs into the database. Once you configure the authorizer property correctly, you can successfully issue the GRANT command.

```
cassandra@cqlsh> grant select permission on keyspace "cycling" to 'test';
```

Since the loss of access to the `system_auth.permissions` table will deny all access to the cluster, ensure that you have several replicas of the system keyspace. If you have multiple datacenters, set the replication class to NetworkTopologyStrategy.

You can grant the following broad types of privileges to a role:

- `SELECT`: Allows a role to read data with the CQL command `SELECT`.

- `MODIFY`: Allows a role to add, modify, and remove data with the CQL commands `INSERT`, `UPDATE`, `DELETE`, and `TRUNCATE`.

- `CREATE`: Grants the ability to create keyspaces and tables with the commands `CREATE KEYSPACE`, and `CREATE TABLE`.

- `ALTER`: Grants the ability to modify keyspaces and tables with the following commands:

  - `ALTER KEYSPACE`

  - `ALTER TABLE`

- `DESCRIBE`: Provides information about objects.

- `DROP`: Enables you to remove objects from the database.

- `EXECUTE`: `SELECT`, `INSERT`, and `UPDATE` privileges using any function, and in any function in a `CREATE AGGREGATE` statement.

- `AUTHORIZE`: Enables the granting and revoking of permissions on keyspaces, tables, functions, and roles.

- `ALL PERMISSIONS`: Enables all types of queries on a table.

The following are examples of how you can issue the `GRANT` command to assign various privileges directly to a role.

You can grant the role *sam* privileges to perform `SELECT` operation on all tables in all keyspaces as follows:

```
cqlsh> grant select on all keyspaces to 'sam';
```

You can grant the role *sam* privileges to perform modifications (`INSERT`, `UPDATE`, `TRUNCATE`, and `DELETE`) on all tables in the CYCLING keyspace as follows:

```
cqlsh> grant modify on keyspace cycling to 'sam';
```

You can grant the role *sam* the ALTER KEYSPACE privilege, as shown here:

```
cqlsh> grant alter keyspace on cycling to 'sam';
```

The ALTER KEYSPACE privilege enables the role *sam* to perform the following operations on the CYCLING keyspace and its tables and indexes:

- ALTER KEYSPACE

- ALTER TABLE

- CREATE INDEX

- DROP INDEX

The ALL PERMISSIONS privilege enables a user to run all types of queries on a table:

```
Cqlsh> grant all permissions on cycling.cyclists to sam;
```

Cassandra's Access Control Matrix

Cassandra employs a hierarchical inheritance system for access control, where the privileges you grant to a resource that's higher in the hierarchy automatically cascade to resources lower in the hierarchy.

For example,

- If you grant a privilege on ALL KEYSPACES, it cascades to all tables in those keyspaces.

- If you grant a privilege on ALL FUNCTIONS, it cascades to all user-defined functions and aggregates.

```
cassandra@cqlsh> use system_auth;
cassandra@cqlsh:system_auth> select * from role_permissions;
 role      | resource     | permissions
-----------+--------------+---------------------------------
      test |         data |                       {'SELECT'}
      test | data/cycling |   {'ALTER', 'MODIFY', 'SELECT'}
 cassandra |   roles/test | {'ALTER', 'AUTHORIZE', 'DROP'}
 cassandra |  roles/test1 | {'ALTER', 'AUTHORIZE', 'DROP'}
(4 rows)
cassandra@cqlsh:system_auth>
```

Configuring Role-Backed Access Control

A database role is a resource to which you grant privileges to access other database resources. Role-based access control is where you assign roles to users, rather than directly assigning privileges to access various database resources.

Typically, in other databases, to use roles to manage access to the database objects, you must do the following, in this order:

- Create a user.

- Create a role.

- Grant privileges on various database resources to the new role.

- Grant the role to the new user.

You follow the same strategy to implement RBAC in a Cassandra database, with the difference that instead of granting roles to users, you grant them to (login) roles (which, as I explained earlier, stand in for users in Cassandra).

In the following sections, I show how to configure RBAC.

Creating Roles for Login Accounts

When you want to grant access to a user or application, you create a role. The following example shows how to create the login role named *manager*:

```
cqlsh> create role manager
        ... with PASSWORD = 'Password123'
        ... and LOGIN = true;
```

By default, the LOGIN property's value is false, meaning that the role won't be able to log into the database. In this case, you want the manager role to be able to log into the database and serve as a login role. You therefore set the LOGIN property to true.

The list roles command shows that the new role was created successfully.

```
cqlsh> list roles;
 role       | super | login | options
-----------+-------+-------+---------
 cassandra |  True |  True |        {}
   manager | False |  True |        {}
      test | False |  True |        {}
     test1 | False |  True |        {}
(4 rows)
cqlsh>
```

You can test the ability of the new account to log in.

```
$ cqlsh 192.168.159.129 -u cassandra -p cassandra
Connected to Test Cluster at 192.168.159.129:9042.
[cqlsh 5.0.1 | Cassandra 3.10 | CQL spec 3.4.4 | Native protocol v4]
Use HELP for help.
cassandra@cqlsh> login manager
Password:
manager@cqlsh>
```

The cqlsh prompt (manager@cqlsh) shows the name of the role, which is *manager* in this example.

You can change the password for a login-enabled role by running the ALTER ROLE command, as shown here:

```
cqlsh>ALTER ROLE manager WITH PASSWORD = 'manager';
cqlsh>
```

Granting Permissions to Roles

You can also create a role and grant various permissions on one or more database objects to that role. You can then assign this role to a login role.

Object Permissions in Cassandra

You can authorize object permissions to authenticated roles on the following database objects:

- Keyspace
- Table
- Function
- Aggregate
- Roles
- MBeans

The following are the types of permissions you can configure on database objects:

- CREATE
- ALTER
- DROP
- SELECT
- MODIFY
- DESCRIBE

Granting Permissions on Objects to Roles

The following example shows how to create a role that you aren't going to use for a login; instead, you'll use the role for granting permissions on database objects. You first create the role *cycling_admin*, and you grant privileges on the keyspace *cycling* to that role. Following this, you grant the role *cycling_admin* to the login role *manager*.

1. Create the role with the CREATE ROLE command.

   ```
   cqlsh> create role cycling_admin;
   ```

   Since this role is for assigning object privileges and not for a login account, you didn't set the SUPERUSER or the LOGIN properties, both of which are False by default.

```
cassandra@cqlsh> list roles;
  role          | super | login | options
----------------+-------+-------+---------
cassandra | True | True |           {}
cycling_admin | False | False |        {}
      manager | False | True |          {}
         test | False | True |          {}
        test1 | False | True |          {}

(5 rows)
cassandra@cqlsh>
```

2. Once you create the new role, grant privileges on a tablespace
    (*cycling*) to that role.

    ```
    cqlsh> grant all permissions on keyspace cycling to cycling_admin;
    ```

3. Grant the new role *cycling_admin* to the login role *manager*,
    which you created in the previous section.

    ```
    cqlsh> grant cycling_admin to manager;
    ```

Listing Permissions

You can list all permissions granted to roles with the list all permissions command.

```
cassandra@cqlsh> list all permissions;
  role          | username       | resource              | permission
----------------+----------------+-----------------------+------------
    cassandra |     cassandra | <role cycling_admin> |     ALTER
    cassandra |     cassandra | <role cycling_admin> |      DROP
    cassandra |     cassandra | <role cycling_admin> | AUTHORIZE
    cassandra |     cassandra |       <role manager> |     ALTER
    cassandra |     cassandra |       <role manager> |      DROP
    cassandra |     cassandra |       <role manager> | AUTHORIZE
    cassandra |     cassandra |          <role test> |     ALTER
    cassandra |     cassandra |          <role test> |      DROP
    cassandra |     cassandra |          <role test> | AUTHORIZE
```

cassandra	cassandra	<role test1>	ALTER
cassandra	cassandra	<role test1>	DROP
cassandra	cassandra	<role test1>	AUTHORIZE
cycling_admin	cycling_admin	<keyspace cycling>	CREATE
cycling_admin	cycling_admin	<keyspace cycling>	ALTER
cycling_admin	cycling_admin	<keyspace cycling>	DROP
cycling_admin	cycling_admin	<keyspace cycling>	SELECT
cycling_admin	cycling_admin	<keyspace cycling>	MODIFY
cycling_admin	cycling_admin	<keyspace cycling>	AUTHORIZE
test	test	<all keyspaces>	SELECT
test	test	<keyspace cycling>	ALTER
test	test	<keyspace cycling>	SELECT
test	test	<keyspace cycling>	MODIFY

```
(22 rows)
cassandra@cqlsh>
```

Check out the Resource column in the query output to see how Cassandra assigns different permissions to the three roles in the database: the default superuser role *cassandra*, the login role *test*, and the new role you created, *cycling_admin*.

Viewing Permissions Granted to Roles

To view the permissions granted to a specific role, run the list all permissions of <role_name> command:

```
cassandra@cqlsh> list all permissions of manager;
```

role	username	resource	permission
cycling_admin	cycling_admin	<keyspace cycling>	CREATE
cycling_admin	cycling_admin	<keyspace cycling>	ALTER
cycling_admin	cycling_admin	<keyspace cycling>	DROP
cycling_admin	cycling_admin	<keyspace cycling>	SELECT
cycling_admin	cycling_admin	<keyspace cycling>	MODIFY
cycling_admin	cycling_admin	<keyspace cycling>	AUTHORIZE

```
(6 rows)
cassandra@cqlsh>
```

Configuring Firewall Ports for Access

Cassandra nodes require several firewall ports to be open so they can communicate. Some of the nodes that should be open are public and others are inter-node or Cassandra-specific ports.

Ensure that the following ports are open:

Type of Port	Port Number	Port Function
Public	22	SSH Port
Cassandra (inter-node)	7000	For inter-node cluster communications
Cassandra (inter-node)	70001	For SSL inter-node communications
Cassandra (inter-node)	7199	JMX monitoring port
Cassandra (client)	9042	Client port
Cassandra (client)	9142	Default for native transport protocol (when you need encrypted and nonencrypted communications)

Encrypting Cassandra with SSL

SSL is a security protocol that encrypts data during communications between Cassandra clients and the nodes, as well as among the nodes. Each entity that participates in communications must have a private key that the entity stores and a public key that it exchanges with other entities. The server sends clients that want to connect securely to a Cassandra cluster a certificate with the public key of the server.

Note   Encrypting data has minimal performance impact, except in the case of a high number of connections.

SSL encryption can secure intra-node communications by encrypting the data flowing between nodes. You can also set up client-node SSL encryption to protect data passed between client programs such as cqlsh or nodetool and a cluster's nodes.

The client validates the security certificate by having the server validate it with its private key. A keystore stores private keys and certificates, and a truststore stores the public keys. A system can use a Certificate Authority (CA), in which case the truststore also stores the CA's signed certificates. The password for a keystore is called the *keypass*, and the password for the truststore is called a *storepass*.

Installing the Java Cryptography Extension Files

The default `cassandra.yaml` file refers to several cipher suites, some of which are available only if you install the *Java Cryptography Extension (JCE) Unlimited Strength Jurisdiction Policy Files* when using Oracle Java with SSL. Installing the JCE files ensures support for all encryption algorithms when you're using Oracle Java. You must install the files on all nodes of the cluster.

Note   Installing the JCE files is a best practice when using Oracle Java with SSL.

Follow these steps to install the JCE files. The steps here show how to install JCE on a Red Hat Linux system.

1. Install the EPEL directory.

   ```
   $ sudo yum install epel-release
   ```

2. Download the JCE files from the Oracle Java SE Download page (www.oracle.com/technetwork/java/javase/downloads/index.html).

3. Unzip the JCE files you've downloaded, and copy the local_policy.jar and the US_export_policy.jar files to the $JAVA_HOME/jre/lib/security directory.

Installing the JCE is even simpler on a Debian based system.

```
$ sudo apt-get install oracle-java8-unlimited-jce-policy
Reading package lists... Done
Building dependency tree
...
Unlimited JCE Policy for Oracle Java 8 installed
$
```

Before you can learn how to work with inter-node and client-node encryption through SSL, you must first set up SSL encryption by preparing SSL server certificates.

Preparing the Server Certificates

For both inter-node and client-node encryption, you must first generate SSL certificates and validate the certificates. Here's what you need to do to get going with SSL certificates.

- Use the `openssl` and `keytool` utilities to generate SSL certificates.

- Generate a self-signed CA to validate the SSL certificates, as I show shortly. You can also have the certificates signed by a trusted public certificate authority such as Verisign.

Note   Nate McCall shows how to set up SSL inter-node encryption using a ccm-based Cassandra cluster. The article is at `http://thelastpickle.com/blog/2015/09/30/hardening-cassandra-step-by-step-part-1-server-to-server.html`.

In this section, I show you how to prepare SSL certificates for a production environment.

Create a Certificate Authority

The first step in encrypting the database with SSL is to create your own CA that you can use to sign all the server-specific certificates. This will help create a trust chain that makes it easy to manage the certificates. Follow these steps to create the CA.

1. The first step is to create the root CA certificate and key. Execute the `openssl req` command and pass this command the certificate configuration file. This requires that you first create a certificate configuration file, which you'll name `my_rootCa_cert.conf` file. Open a vi or nano editor and enter the following information in the `my_rootCa_cert.conf` file:

```
$ vi my_rootCa_cert.conf
# my_rootCa_cert.conf
[ req ]
```

```
distinguished_name      = req_distinguished_name
prompt                  = no
output_password         = myPass
default_bits            = 2048

[ req_distinguished_name ]
C                       = US
O                       = MyCompany
OU                      = TestCluster
CN                      = rootCa
$
```

2. Once you create the certificate configuration file (my_rootCa_
    cert.conf), run the openssl req command as shown here,
    making sure that you pass the configuration file name via the
    config attribute.

    ```
    $ sudo openssl req -config my_rootCa_cert.conf  -new -x509
    -nodes -subj /CN=rootCa/OU=TestCluster/O=YourCompany/C=US/
    -keyout rootCa.key -out rootCa.crt -days 365
    Generating a 2048 bit RSA private key
    ...............................+++
    ..............................................................
    .....+++
    writing new private key to 'rootCa.key'
    $
    ```

    This command will create the files rootCa.key and rootCa.crt,
    which you'll use later. The rootCa.key file is the file to which you'll
    write the key and the rootCa.cert file is the file to which you'll
    write the certificate.

3. Verify the rootCa certificate (the rootCa.crt file) with the openssl
    x509 command.

    ```
    $ sudo openssl x509 -in rootCa.crt -text -noout
    Certificate:
        Data:
    ```

```
       Version: 1 (0x0)
       Serial Number: 12633404970202873071
       (0xaf52dee2c098bcef)
   Signature Algorithm: sha256WithRSAEncryption
       Issuer: C=US, O=YourCompany, OU=TestCluster,
       CN=rootCa
       Validity
           Not Before: Aug 25 19:00:47 2017 GMT
           Not After : Aug 25 19:00:47 2018 GMT
 ...
 $
```

Creating Certificates for All Nodes

The next step is to generate the public and private key pairs using the keytool utility. You must do this on all nodes. You can generate the certificates on one node and copy them over to all the nodes later.

```
$ sudo keytool -genkeypair
-keyalg RSA -alias 192.168.159.129
-keystore 192.168.159.129.jks
-storepass myKeyPass
-keypass myKeyPass
-validity 365
-keysize 2048
-dname "CN=192.168.1159.129, OU=TestCluster, O=YourCompany, C=US"
$
```

In this example, you have two nodes in the test cluster, so you need to run this command on both. The keytool command shown here is for the node with the IP address 192.169.159.129. For the second node in the cluster, which has an IP address of 192.168.159.130, you run the same command, but you replace the IP address 192.168.159.129 with the IP address 192.168.159.130.

In this command,

- -genkeypair generates a public/private key pair.

- -keystore species the keystore filename (you use the IP address so as to map the files to the Cassandra nodes).

- -storepass specifies the keystore password.

- -keypass specifies the private key password (must be the same as the value for the -storepass attribute).

- -dname specifies the Distinguished Name (DN) to be associated with the value of alias. The CN value is set to the node's IP address or FQDN.

Check the certificates you've generated to ensure that the key store is accessible and that it contains the correct key pair.

```
$sudo keytool -list -keystore 192.168.159.129.jks -storepass myKeyPass

Keystore type: JKS
Keystore provider: SUN

Your keystore contains 1 entry

192.168.159.129, Aug 25, 2017, PrivateKeyEntry,
Certificate fingerprint (SHA1): C2:C1:6C:F5:DF:F7:EC:D6:09:66:BB:67:17:38:5
8:87:38:E1:AE:DD
$
```

Exporting the Certificate Signing Requests

Once you generate the node certificate and the key, you must export a certificate signing request (CSR) for each node. The CSR is signed with the rootCa certificate to verify that the certificate is trusted.

```
$ sudo  keytool -certreq
-keystore 192.168.159.129.jks
-alias 192.168.159.129
-file 192.168.159.129.csr
-keypass myKeyPass -storepass myKeyPass
-dname "CN=192.168.1159.129, OU=TestCluster, O=YourCompany, C=US"
$
```

The keytool -certreq command exports a certificate for signing by a CA.

You must repeat this command for each node in the cluster by replacing the IP address shown here (192.168.159.129) with the IP address of the other nodes.

Signing the Certificates with the CA's Public Key

Next, sign the node certificate with the rootCa for each node in the Cassandra cluster, through OpenSSL.

```
$ sudo openssl x509
-req
-CA rootCa.crt
-CAkey rootCa.key
-in 192.168.159.129.csr
-out 192.168.159.129.crt_signed
-days 365
-CAcreateserial -passin pass:myPass

Signature ok
subject=/C=US/O=YourCompany/OU=TestCluster/CN=192.168.1159.129
Getting CA Private Key
$
```

Once again, you must run the openssl command shown here for each of your cluster's nodes.

In this command,

- -CA identifies the rootCa certificate.

- -Cakey identifies the rootCa key.

- -in specifies the filename from which to read the certificate.

- -out specifies the output filename for the signed certificate.

Use the rootCa certificate and the signed certificate to verify the signed certificate.

```
$ openssl verify -CAfile rootCa.crt 192.168.159.129.crt_signed
192.168.159.129.crt_signed: OK
$
```

Adding the CA to the Keystore

Import the rootCa certificate to each node's keystore.

```
$ sudo keytool -importcert
    -keystore 192.168.159.129.jks
    -alias rootCa
    -file rootCa.crt
    -noprompt
    -keypass myKeyPass
    -storepass myKeyPass

Certificate was added to keystore
$
```

The keystore file will now have entries for both the rootCa certificate and the node certificate.

Importing the Signed Certificates into the Keystore

Each of the key stores on the nodes now contains the CA. The next step is to do the same thing as the previous step, but this time import the node's signed certificate into the node keystore for each node.

```
$ sudo keytool -importcert
-keystore 192.168.159.129.jks
-alias 192.168.159.129
-file 192.168.159.129.crt_signed
-noprompt
-keypass myKeyPass
-storepass myKeyPass

Certificate reply was installed in keystore
$
```

The node certificate you created earlier is replaced with the signed node certificate.

Building the Trust Store

You must next create a server truststore that will establish a chain of trust between the cluster's nodes by verifying connection requests from other nodes. You run the keytool -importcert command to build the key store.

```
$ keytool -importcert
    -keystore generic-server-truststore.jks
    -alias rootCa
    -file rootCa.crt
    -noprompt
    -keypass myPass
    -storepass truststorePass

Certificate was added to keystore
$
```

Once you create the trust store, you can share it across your cluster. The trust store says that it'll trust connections from all nodes whose client certificates were signed by this CA.

You can verify the truststore file thus:

```
$ sudo keytool -list -keystore generic-server-truststore.jks -storepass
truststorePass

Keystore type: JKS
Keystore provider: SUN

Your keystore contains 1 entry

rootca, Aug 28, 2017, trustedCertEntry,
Certificate fingerprint (SHA1): BB:BF:D7:8F:15:4E:41:91:37:70:EB:4C:67:AB:2
C:25:37:A4:18:B0
$
```

Configuring the Cluster with the Keystores

Earlier, I showed how to create both the node-specific key stores and a generic trust store. Now you must move them to where Cassandra can find them. Cassandra looks for the key stores and trust stores in the $CASSANDRA_HOME/conf directory by default.

First, you copy over the node-specific key stores for each of the two nodes in the cluster.

```
cp 192.168.192.159.129.jks  $CASSANDRA_HOME/conf/server-keystore.jks
cp 192.168.192.159.130.jks  $CASSANDRA_HOME/conf/server-keystore.jks
```

Next, you copy over the generic trust store to both nodes.

```
cp generic-server-truststore.jks  $CASSANDRA_HOME/conf/server-truststore.jks
cp generic-server-truststore.jks  $CASSANDRA_HOME/conf/server-truststore.jks
```

At this point, the configuration of SSL certificates is complete. You must now enable encryption.

Enabling Inter-Node Encryption

The last step to encrypt inter-node communications is to enable encryption by modifying the cassandra.yaml file. In the cassandra.yaml file, modify the server_ encryption_options section as follows:

```
server_encryption_options:
    internode_encryption: all                /default value: none
    keystore: /cassandra/conf/server-keystore.jks
    keystore_password: cassandra
    truststore: /cassandra/conf/server-truststore.jks
    truststore_password: cassandra
    protocol: TLS
    algorithm: SunX509
    store_type: JKS
    # cipher_suites: [TLS_RSA_WITH_AES_128_CBC_SHA,TLS_RSA_WITH_AES_256_
    CBC_SHA,TLS_DHE_RSA_WITH_AES_128_CBC_SHA,TLS_DHE_RSA_WITH_AES_256_CBC_
    SHA,TLS_ECDHE_RSA_WITH_AES_128_CBC_SHA,TLS_ECDHE_RSA_WITH_AES_256_CBC_
    SHA]
    require_client_auth: true
    require_endpoint_verification: false
```

Earlier, I showed how to install strong encryption policy files in in the JDK (JCE Policy Files). This supports the use of strong cyphers such as 256bit AES, as shown by the cipher_suites property in the cassandra.yaml file.

You're finally all set for encryption. Restart the cluster.

```
$ kill -9 cassandra-pid
$ $CASSANDRA_HOME/bin/cassandra
INFO  [main] 2017-08-28 12:33:50,572 MessagingService.java:687 - Starting
Encrypted Messaging Service on SSL port 7001
```

The message "Starting Encrypted Messaging Service on SSL port 7001" shows that SSL encryption is in place for inter-node communications.

Enabling Client Encryption

In the previous section, I showed how to configure node-to-node encryption. However, communications between clients such as cqlsh and the cluster will still be unencrypted. In this section, I show how to configure client-node encryption. You use the same SSL certificates that you created for inter-node communications. You just need to enable client encryption and add the CA certificates to the cqlshrc file.

To enable client encryption, here's how you modify the cassandra.yaml file:

```
# enable or disable client/server encryption.
client_encryption_options:
    enabled: true
    # If enabled and optional is set to true encrypted and unencrypted
    connections are handled.
    optional: false
    keystore: /cassandra/apache-cassandra-3.10/conf/server-keystore.jks
    keystore_password: myKeyPass
     require_client_auth: true
    # Set trustore and truststore_password if require_client_auth is true
     truststore: /cassandra/apache-cassandra-3.10/conf/server-truststore.jks
     truststore_password: truststorePass
    # More advanced defaults below:
     protocol: TLS
     algorithm: SunX509
     store_type: JKS
     cipher_suites: [TLS_RSA_WITH_AES_128_CBC_SHA,TLS_RSA_WITH_AES_256_CBC_
    SHA,TLS_DHE_RSA_WITH_AES_128_CBC_SHA,TLS_DHE_RSA_WITH_AES_256_CBC_SHA,TLS_
    ECDHE_RSA_WITH_AES_128_CBC_SHA,TLS_ECDHE_RSA_WITH_AES_256_CBC_SHA]
```

You must restart the cluster at this point.

In the logs generated during the node startup, you'll see the following message, indicating that SSL encryption is now enabled between clients and the database server:

```
INFO  [main] 2017-09-05 14:17:10,681 Server.java:145 - Enabling encrypted
CQL connections between client and server
```

Once the cluster comes back up, since SSL validation is enabled now, you must specify the --ssl option to connect to Cassandra:

```
$ cqlsh --ssl
Validation is enabled; SSL transport factory requires a valid certfile
to be specified. Please provide path to the certfile in [ssl] section as
'certfile' option in /home/cassandra/.cassandra/cqlshrc (or use [certfiles]
section) or set SSL_CERTFILE environment variable.
$
```

The reason you see this error is because you have enabled SSL for client communications and hence cqlsh, which is a client, doesn't trust the cassandra node now. For it to trust the node, it should see the CA certificate that you generated earlier (rootCa.crt). This ensures that Cassandra is using a certificate that was signed by the CA. You point to the CA certificate by listing it under the [ssl] section of the cqlshrc file (/home/cassandra/.cassandra/cqlshrc), as shown here:

```
[connection]
hostname=192.168.159.130
port=9042
factory = cqlshlib.ssl.ssl_transport_factory

[ssl]
certfile = /cassandra/apache-cassandra-3.10/conf/rootCa.crt
validate = true
```

Now you'll be able to connect to Cassandra through cqlsh using SSL.

```
$ cqlsh -ssh
```

You now have a secure connection to the Cassandra node through the cqlsh client.

JMX Authentication and Authorization

JMX authentication and authorization allows you to control user access to JMX-based tools such as nodetool and JConsole. You can configure JMX connections to use Cassandra's internal authentication and authorization mechanisms, just as the CQL clients do.

If you've configured usernames and passwords in the database through the authentication and authorization mechanism described earlier in this chapter, you must execute JMS tools as well with the authentication and authorization you've configured.

Before you can use nodetool or JConsole with authentication, you must enable JMX authentication and authorization.

Enabling JMX Authentication and Authorization

By default, JMX security isn't configured, meaning that you can access JMS tools such as nodetool and JConsole only from the localhost. You can configure JMX authentication and authorization using local password and access files for configuring the credentials for users and access permission. However, in Cassandra 3.6 onwards, you can also configure JMX security by piggybacking on Cassandra's internal authentication and authorization.

In this section, I show how to configure JMX authentication and authorization using local files.

Here are the steps to configure JMX security using local files.

1. Edit the `cassandra-env` file (in the `$CASSANDRA_HOME/conf` directory), and change the two settings in the chunk shown here:

    ```
    if [ "$LOCAL_JMX" = "yes" ]; then
      JVM_OPTS="$JVM_OPTS -Dcassandra.jmx.local.port=$JMX_
      PORT"
      JVM_OPTS="$JVM_OPTS -Dcom.sun.management.jmxremote.
      authenticate=false"
    else
    ```

    In the first line, change the "yes" to "no". In the third line, change "false" to "true".

2. Create the password file /etc/cassandra/jmxremote.password
   (the default location) and add the following line to the file:

   ```
   cassandra   cassandra
   ```

   I'm using the default superuser account here, but this isn't safe in
   a production system, and you must replace this set of credentials
   with one or more credentials for all users that you'd like to be able
   to access JMX-compliant utilities such as nodetool.

3. Create the access file (by default) /etc/cassandra/jmxxremote.
   access and add the following information to the access file:

   ```
   cassandra readwrite
   create javax.management.monitor.,javax.management.timer.
   \unregister
   ```

   The readwrite privilege you gave to the role *cassandra* enables
   this JMX client to work with MBeans. More specifically, it
   allows this client to set attributes, invoke operations, receive
   notifications, and so on.

4. Restart the database.

5. Try running the nodetool status command. Since you configured
   JMX authorization and authentication, you shouldn't be able
   to run this command as usual. If you've configured everything
   correctly, you should instead see the following error:

   ```
   $ nodetool status
   nodetool: Failed to connect to '127.0.0.1:7199' -
   SecurityException: 'Authentication failed! Credentials required'.
   ```

6. Run the nodetool status command by providing the credentials
   for the role *cassandra*.

   ```
   $ nodetool -u cassandra -pw cassandra status
   Datacenter: datacenter1
   =======================
   Status=Up/Down
   ```

```
|/ State=Normal/Leaving/Joining/Moving
--  Address          Load        Tokens        Owns
(effective)  Host ID                            Rack
UN  192.168.159.129  24.21 MiB   256           49.9%
0dbb9e0e-867e-4179-b6b6-631d38dd68f9  rack1
UN  192.168.159.130  24.25 MiB   256           50.1%
001399d4-49fc-467c-b188-e93629a0f118
```

The database is now configured for JMX authentication and authorization. As with the nodetool utility, you'll now be required to provide a username and password when you use JConsole to remotely connect to a Cassandra cluster.

Using cqlsh with Authentication

You can configure authentication so that logging into cqlsh requires a password. You do this by creating or modifying the `cqlshrc` file. The following are the steps to configure this authentication.

1. Edit the cqlshrc file, or if you don't have one, create one in the following location:

    ```
    /etc/cassandra/cqlshrc.sample
    /* package installation
    install_location/conf/cqlshrc.sample
    /* tarball installation
    ```

2. Enter the following lines in the `cqlshrc` file:

    ```
    [authentication]
    username = sam
    password = !!bang!!$
    ```

3. Secure the file by setting the following permissions:

    ```
    $ sudo chmod 400 home/.cassandra/cqlshrc
    ```

Summary

Securing Cassandra is a multi-faceted affair. To secure your data comprehensively, besides configuring authentication, authorization, and SSL encryption of inter-cluster and client-cluster communications, you may also want to encrypt data at rest.

Several good third-party encrypting solutions exist. By using one or more them you can complete the circle and reduce the chances of a security breach. I didn't discuss these products in this chapter, but a quick search on the Internet for "encryption at rest" should show you the available choices.

Index

A

© Sam R. Alapati 2018
S. R. Alapati, *Expert Apache Cassandra Administration*, https://doi.org/10.1007/978-1-4842-3126-5

C

Get the eBook for only $5!

Why limit yourself?

With most of our titles available in both PDF and ePUB format, you can access your content wherever and however you wish—on your PC, phone, tablet, or reader.

Since you've purchased this print book, we are happy to offer you the eBook for just $5.

To learn more, go to http://www.apress.com/companion or contact support@apress.com.

Apress®

All Apress eBooks are subject to copyright. All rights are reserved by the Publisher, whether the whole or part of the material is concerned, specifically the rights of translation, reprinting, reuse of illustrations, recitation, broadcasting, reproduction on microfilms or in any other physical way, and transmission or information storage and retrieval, electronic adaptation, computer software, or by similar or dissimilar methodology now known or hereafter developed. Exempted from this legal reservation are brief excerpts in connection with reviews or scholarly analysis or material supplied specifically for the purpose of being entered and executed on a computer system, for exclusive use by the purchaser of the work. Duplication of this publication or parts thereof is permitted only under the provisions of the Copyright Law of the Publisher's location, in its current version, and permission for use must always be obtained from Springer. Permissions for use may be obtained through RightsLink at the Copyright Clearance Center. Violations are liable to prosecution under the respective Copyright Law.

Printed in the United States
By Bookmasters